MODFLOW-LGR—Documentation of Ghost Node Local Grid Refinement (LGR2) for Multiple Areas and the Boundary Flow and Head (BFH2) Package

By Steffen W. Mehl and Mary C. Hill

Prepared in cooperation with the U.S. Department of Energy
A Product of the Groundwater Resources Program

Techniques and Methods 6–A44

U.S. Department of the Interior
U.S. Geological Survey

U.S. Department of the Interior
SALLY JEWELL, Secretary

U.S. Geological Survey
Suzette M. Kimball, Acting Director

U.S. Geological Survey, Reston, Virginia: 2013

For more information on the USGS—the Federal source for science about the Earth, its natural and living resources, natural hazards, and the environment, visit http://www.usgs.gov or call 1–888–ASK–USGS.

For an overview of USGS information products, including maps, imagery, and publications, visit http://www.usgs.gov/pubprod

To order this and other USGS information products, visit http://store.usgs.gov

Suggested citation:
Mehl, S.W., and Hill, M.C., 2013, MODFLOW–LGR—Documentation of ghost node local grid refinement (LGR2) for multiple areas and the boundary flow and head (BFH2) package: U.S. Geological Survey Techniques and Methods book 6, chap. A44, 43 p., http://pubs.usgs.gov/tm/6A44/.

Preface

This report describes ghost node Local Grid Refinement (LGR2) for MODFLOW-2005, the U.S. Geological Survey's three-dimensional finite-difference groundwater model. LGR2 is designed to allow users to create MODFLOW simulations using one or more refined grids that are embedded within a coarser grid.

This report also describes the Boundary Flow and Head (BFH2) Package for MOD-FLOW-2005. After simulating the coupled models with LGR2, the BFH2 Package allows each refined grid or the parent grid to be run independently from the previously coupled model grid(s).

The performance of the programs has been tested in a variety of applications. Future applications, however, might reveal errors that were not detected in the test simulations. Users are requested to notify the U.S. Geological Survey of any errors found in this document or the computer program using the e-mail address available at the web address below. Updates might occasionally be made to both this document and to LGR2 and the BFH2 Package. Users can check for updates on the Internet at URL *http://water.usgs.gov/software/ground_water.html/*.

Contents

Abstract..1

Introduction...1

    Purpose and Scope 2

    Acknowledgments 2

Highlights and Compatibility..2

    Highlights for New Users and Quick Reference...2

        Accuracy...2

        Execution Time 3

        Model Setup...3

        Grid and Time-Step Design...3

        Multiple Areas of Refinement...3

    Compatibility with Other MODFLOW Packages...4

        Running Parent and Child Models Independently Using the Boundary Flow and
            Head Package..4

        Using LGR2 to Simulate Solute Transport and Particle Tracking 5

Description of Local Grid Refinement (LGR)..5

    Grid of the Ghost-Node Method—Parent Grid, Child Grid, and the Interface 5

        Lateral Interface between the Parent and Child Grids..5

        The Top and Bottom of the Child Grid and Vertical Refinement..8

    Including Multiple Areas of Refinement 9

    The Iterative Coupling 9

        Ghost-Node Coupling 10

        Two Methods of Formulating and Solving the Equations...11

        Determining Ghost Node Specified Heads...12

            Interpolation Concepts Illustrated Analytically Using a Two-Dimensional
               Model 12

            Darcy-Planar Interpolation 13

        Unconfined Conditions...13

    Closure Criteria...14

        Closure Criteria for LGR Iterations 14

        Solver Iterations...14

    Transient Simulations..14

Examples 15

    Example 1: Two-Dimensional Steady State Test Case with Heterogeneity and Pumping......15

        Calculation of Heads 15

            Effect of Heterogeneity Contrast...17

    Example 2: Three-Dimensional, Unconfined, Transient Test Case with Heterogeneity,
        Pumping, and Rewetting 18

    Example 3: Three-Dimensional Steady State Test Case with Homogeneity,
        Stream-Aquifer Interactions 20

        Convergence and Analysis of Flux Errors..20

    Example 4: Multiple Refined Areas 22

References Cited...23

Appendix 1. LGR2 Input Instructions and Selected Input and Output Files from Examples 1
and 3 26
LGR2 Input Instructions 26
Explanation of Variables Read by LGR2 27
Example LGR2 Input Files...28
Sample LGR2 Output...29
Reference Cited..30
Appendix 2. Independent Simulations Using the Boundary Flow and Head (BFH2) Package........31
Example BFH2 Inputs..31
Sample BFH2 Output...32
Reference Cited..33
Appendix 3. Error Propagation in LGR2 34
Reference Cited..35
Appendix 4. LGR2 Input File for Multiple Refined Areas 36
LGR2 Input File Showing Available Flexibility...36
Appendix 5. Brief Program Description 38
Variables in Fortran Module LGRMODULE 38
Description of LGR2 Subroutines 40
Variables in Fortran Module GWFBFHMODULE 41
Description of BFH2 Subroutines..42
Appendix 6. Relative Advantages of Ghost-Node versus Shared-Node Coupling 43
Advantages of the Ghost-Node Method 43
Advantages of the Shared-Node Method 43

Figures

1. Flow chart for the iteratively coupled Local Grid Refinement procedure with
multiple child grids. 6
2. Flowchart of MODFLOW-2005 with Local Grid Refinement.......................................7
3. Two-dimensional areal schematic through the center of the locally refined grid............8
4. Cross-sectional schematic of vertical refinement interface of (a) a one-layer
parent model refined to a three-layer child model, (b) a multi-layer parent model
where the child refinement varies vertically using both even and odd refinement
ratios and terminates at the bottom of the second parent layer, (c) a multi-layer
parent model where the child refinement varies vertically using a 1:1 refinement
ratio in the first layer and a 3:1 refinement ratio in subsequent layers. The
refinement extends to the bottom of the parent model, and (d) a multi-layer
parent model with a single layer child model, which is not possible 9
5. Schematic of a coarse grid with two areas of local grid refinement separated
by two parent grid cells 9
6. Schematic of a 2:1 vertically and horizontally refined grid interface showing
shaded prisms between the ghost node and adjacent child node that are used
in the calculation of ghost node conductances. 11

7. Darcy-planar interpolation in relation to linear interpolation between ghost nodes for cells with different hydraulic conductivity, as denoted by the different shading of the two cells...12

8. Heterogeneity structure and area of local refinement around the well indicated by dashed rectangle..16

9. Flow vectors and head contours calculated using the low-contrast set of transmissivities listed in figure 8..16

10. Flow vectors and head contours calculated using the high-contrast set of transmissivities listed in figure 8..17

11. Simulated hydraulic heads of the parent model showing drying and rewetting of cells..19

12. Plan view of a three-dimensional aquifer system used to test the local grid refinement method..20

13. Total river leakage in the child model as simulated using the globally refined and locally refined models in relation to number of iterations of the local grid refinement procedure are shown with dashed lines...21

14. Example of grid refinement around two areas of pumping within a coarse-grid model...22

Tables

1. Compatibility issues and possible adjustments required by the user for MODFLOW-2005 Groundwater Flow Process Packages..4

2. Comparison of errors and computer processing time for the shared-node and ghost-node grid refinement methods applied to the low-contrast version of example 1..17

3. Comparison of errors and computer processing time for the shared-node and ghost-node grid refinement methods applied to the high-contrast version of example 1..18

4. Average percent difference in river leakages in the child model using a one-way coupled and iteratively coupled solutions compared to the globally refined solution..21

5. Comparison of simulated hydraulic head at the wells and computer processing time for several grids. Only one well is reported because the flow field is symmetric, as shown in figure 14..22

5-1. Variables in Fortran module LGRMODULE...38

5-2. Variable in Fortran module GWFBFHMODULE..41

Conversion Factors

Multiply	By	To obtain
	Length	
meter (m)	3.281	foot
meter per day (m/day)	3.281	foot per day
	Flow rate	
square meter per second (m²/s)	10.76	square foot per day
square meter per hour (m²/hr)	10.76	square foot per hour
cubic meter per second (m³/s)	35.31	cubic foot per second
cubic meter per hour (m³/hr)	35.31	cubic foot per hour

Acronyms

MODFLOW-2005 Packages and Capabilities

BCF	Block-Centered Flow
BFH	Boundary Flow and Head
CHD	Constant-Head Boundary
DE4	Direct solution based on alternating diagonal ordering
DIS	Discretization
DRN	Drain
DRT	Drain Return Flow
EVT	Evapotranspiration
FHB	Flow and Head Boundary
GHB	General-Head Boundary
GMG	Geometric Multigrid (solver)
HFB	Horizontal-Flow Barrier
HUF	Hydrologic-Unit Flow
LGR	Local Grid Refinement
LMG	Link Algebraic Multigrid (solver available freely to USGS personnel)
LPF	Layer Property Flow
LVDA	Layer Variable-Direction Horizontal Anisotropy
MNW	Multi-Node Well
PCG	Preconditioned Conjugate-Gradient (solver)
RCH	Recharge
RIV	River
SIP	Strongly Implicit Procedure (solver)
SFR	Streamflow Routing—new version
STR	Streamflow Routing—old version

TMR   Telescopic Mesh Refinement
WEL   Well

MODFLOW-LGR—Documentation of Ghost Node Local Grid Refinement (LGR2) for Multiple Areas and the Boundary Flow and Head (BFH2) Package

By Steffen W. Mehl and Mary C. Hill

Abstract

This report documents the addition of ghost node Local Grid Refinement (LGR2) to MODFLOW-2005, the U.S. Geological Survey modular, transient, three-dimensional, finite-difference groundwater flow model. LGR2 provides the capability to simulate groundwater flow using multiple block-shaped higher-resolution local grids (a child model) within a coarser-grid parent model. LGR2 accomplishes this by iteratively coupling separate MODFLOW-2005 models such that heads and fluxes are balanced across the grid-refinement interface boundary. LGR2 can be used in two-and three-dimensional, steady-state and transient simulations and for simulations of confined and unconfined groundwater systems.

Traditional one-way coupled telescopic mesh refinement methods can have large, often undetected, inconsistencies in heads and fluxes across the interface between two model grids. The iteratively coupled ghost-node method of LGR2 provides a more rigorous coupling in which the solution accuracy is controlled by convergence criteria defined by the user. In realistic problems, this can result in substantially more accurate solutions and require an increase in computer processing time. The rigorous coupling enables sensitivity analysis, parameter estimation, and uncertainty analysis that reflects conditions in both model grids.

This report describes the method used by LGR2, evaluates accuracy and performance for two-and three-dimensional test cases, provides input instructions, and lists selected input and output files for an example problem. It also presents the Boundary Flow and Head (BFH2) Package, which allows the child and parent models to be simulated independently using the boundary conditions obtained through the iterative process of LGR2.

Introduction

Simulations of groundwater flow and transport often need highly refined grids in local areas of interest to improve simulation accuracy. For example, refined grids may be needed in (1) regions where hydraulic gradients change substantially over short distances, as would be common near pumping or injecting wells, rivers, drains, and focused recharge; (2) regions of site-scale contamination within a regional aquifer where simulations of plume movement are of interest; and (3) regions requiring detailed representation of heterogeneity, as may be required to simulate faults, lithologic displacements caused by faulting, fractures, thin lenses, pinch outs of geologic units, and so on. Refinement of the finite-difference grid used by MODFLOW can be achieved using globally refined grids, variably spaced grids, or locally refined grids.

Using a globally refined grid (a grid refined over the entire domain) can be computationally intensive. In some cases, the execution times are so long that they result in a computationally intractable problem; in other cases, unnecessarily long execution times interfere with model development and utility. In addition, it may be inconvenient and unnecessarily labor intensive to develop the data sets required to refine an entire grid when only a local area is of interest.

Using a variably spaced grid, a fine grid can be attained locally with more moderate increases in computational time, but often results in refinement in areas that do not need such detail. This arises because the finite-difference method requires that the same grid spacing extend out to the boundaries and has two important implications: (1) if refinement is needed in multiple areas of the domain, using a variably spaced grid often results in a relatively fine grid over the entire domain, and (2) in addition to introducing surplus nodes and therefore more computations, this approach can produce finite-difference cells with a large aspect ratio, which can lead to numerical errors (de Marsily, 1986, p. 351).

Using a locally refined grid can be less computationally intensive than the other two methods. The method described in this report (ghost node Local Grid Refinement in LGR version 2.0; LGR2), links two or more different-sized finite-difference grids: a coarse grid covering a large area which incorporates regional boundary conditions, and a fine grid covering a smaller area of interest. These grids are called parent and child grids, respectively, in this report. Grid refinement can be vertical as well as horizontal, and can span the entire top layer of the parent grid. The link between the parent and child grids can be accomplished as a so-called one-way coupling, in which conditions simulated by the parent grid are imposed on the boundary of the child grid. Alternatively, the link can be accomplished in a way that also includes feedback from the child grid to the parent grid, thus allowing two-way communication between the grids. Solutions with feedback can be achieved through iteration of parent and child solutions until an agreement of heads and fluxes is achieved at the interface of the model grids.

In the field of groundwater modeling, one-way coupling is commonly called telescopic mesh refinement (TMR) and is most commonly accomplished using some form of interpolation of heads, fluxes, or both, from the coarse grid onto the boundaries of the child grid (for example, Ward and others, 1987; Leake and Claar, 1999; Davison and Lerner, 2000; Hunt and others, 2001). This approach is fairly straightforward and works well for many problems. However, the one-way coupling does not allow for feedback from the child grid to the parent grid. Thus, after running both models, the burden is placed on the modeler to check if heads along and fluxes across the interfacing boundary are consistent for both models (Leake and Claar, 1999, p. 5–7). If they do not match, there is no formal mechanism for adjusting the models to achieve better agreement. In this way, TMR methods generally lack numerical rigor and are prone to significant, often undetected errors (Mehl and Hill, 2005).

A numerically rigorous method that ensures that heads and fluxes are consistent between the two grids is needed to obtain dependably accurate solutions. LGR2, as documented in this report, can use a two-way iterative coupling to ensure that the models have consistent boundary conditions along their adjoining interface. The method implemented here couples the models using ghost nodes, rather than shared nodes as described in the original version of MODFLOW-LGR (Mehl and Hill, 2005). In the ghost node-method, grids are constructed such that cell faces of the parent grid are coincident with selected cell faces of the child grid and ghost nodes are used to provide the coupling boundary conditions.

Purpose and Scope

The purpose of this report is to document LGR2 for MODFLOW-2005 (Harbaugh, 2005), which replaces the original version of MODFLOW-LGR. This report first provides highlights of LGR2 and discusses its compatibilities with other MODFLOW-2005 capabilities. The ghost node-method of local grid refinement used by LGR2 is described in detail, and its performance is compared to the shared node coupling of Local Grid Refinement (LGR) for simple two- and three-dimensional problems. Next, input instructions and selected input and output files are provided in Appendixes 1 and 2 for LGR2 and the Boundary Flow and Head Package (BFH2), respectively. Error propagation in LGR2 is described in Appendix 3. Example input for multiple refined areas is presented in Appendix 4. Implementation details of the Fortran

program are provided in Appendix 5. Finally, the relative advantages of the ghost-node coupling of LGR2 are compared to the shared node coupling of LGR in Appendix 6.

Acknowledgments

LGR2 was developed with support from the USGS Groundwater Resources Program and the U.S. Department of Energy. Arlen Harbaugh designed the data storage conventions of MODFLOW-2005 (Harbaugh, 2005) to allow for multiple grids. His careful organization was instrumental for making LGR2 possible and became even more important with the addition of multiple areas of refinement. We are grateful for the code testing, reviews, and comments on this and previous LGR reports provided by Chuck Heywood and Kyle Richards.

Highlights and Compatibility

This section presents highlights important to users deciding on whether to use LGR2 versus an alternative method of grid refinement. Relative advantages of the ghost node-method of LGR2 are compared to the relative advantages of the shared node method of LGR. User concerns for designing models that are compatible with LGR2 are listed here for user convenience; additional discussion of these points is presented in the report, as noted.

Highlights for New Users and Quick Reference

Highlights of LGR are organized into five topics: (1) accuracy, (2) execution time, (3) model setup, (4) grid and time step design, and (5) multiple areas of refinement. LGR2 can be run to perform one-way coupling by setting MXLGRITER=1; the comments here apply when LGR2 is run iteratively (MXLGRITER > 1). MXLGRITER is defined in Appendix 1.

Accuracy

1. Local refinement can provide much of the improved accuracy achievable by global refinement with much smaller execution times (see the Example 1: Two-Dimensional Steady State Test Case with Heterogeneity and Pumping section of this report, and figure 26 of Mehl and Hill, 2005).

2. Local refinement generally improves the accuracy of all parts of the simulated system. (see "Parent Grid Error" and "Child Grid Error" sections in Mehl and Hill, 2005).

3. The greatest refinement ratio does not necessarily produce the most accurate solution (see "Effects of the Refinement Ratio" section in Mehl and Hill, 2005).

4. Local grid refinement maintains the rate of convergence of globally refined grids for homogeneous and heterogeneous models. This means locally refined grids reduce error in a way that is consistent with global refinement and supports local refinement as a valid alternative to global refinement (see "Convergence Properties" section in Mehl and Hill, 2005).

Execution Time

1. LGR uses a solution method that iterates between the parent model and the child models. A single iteration requires one parent-model solution (execution time T_{parent}) and one child-model solution for each child model used (execution time T_{child}). The execution time per iteration is approximately the execution time of the parent grid, T_{parent}, plus the sum of the execution times for each child, ΣT_{child}. The number of iterations varies depending on the flow system, heterogeneity, and the grid discretization. Generally, between 10 to 20 iterations are sufficient for most problems (see results in table 2 and table 3 and example 3 of this report).

Model Setup

1. The parent and child models each require a MODFLOW Name file (Harbaugh and others, 2000, p. 7, 43) and associated set of input files. The unit numbers defined in these files need to be unique—a unit number used in the parent-grid model input and output files cannot be used for the child-grid model. For simulations involving multiple child models, the unit numbers must be unique to each child model.

2. In the Basic Package input file (Harbaugh and others, 2000, p. 50) for the child model, set IBOUND = IBFLG (see Input Instructions in Appendix 1) for cells that border the parent model. This is the perimeter of the child model. Except for the perimeter of the child model, do not use the values defined by IBFLG and -IBFLG anywhere in the IBOUND arrays of the child or parent models.

3. LGR2 currently (2013) needs to have sensitivities calculated and parameter estimation performed using universally applicable programs such as UCODE_2005 (Poeter and others, 2005), PEST (Doherty, 2010), or OSTRICH (Matott, 2005).

Grid and Time-Step Design

1. For LGR2 currently (2013), only block-shaped volumes of local refinement can be simulated.

2. The ghost-node coupling used by LGR2 requires child-grid spacing that is an odd or even integer factor of the parent grid. For example, ratios of refinement of 1:1, 2:1, 3:1, 4:1, 5:1, 6:1, 7:1, and so on can be simulated. Fractional ratios such as 5:1 cannot be simulated.

3. For vertical refinement, the top of the child grid needs to coincide with the top of the parent grid. However, vertical grid refinement does not need to start at the top because a vertical refinement ratio of 1:1 can be used. This can be useful when thick upper layers are desired for simulating water-table conditions (see "The Top and Bottom of the Child Grid and Vertical Refinement" section). Furthermore, the entire top layer of a parent model can be refined with a child grid. This may be useful in situations where detail is needed at the upper surface of the model (for example, to simulate stream-aquifer interactions), but is not needed in deeper parts of the system.

4. For transient simulations, the time-step size needs to be the same for both models. This is not a limitation of the method, but of the implementation currently (2013) used. It is most easily accomplished by defining identical stress period lengths and time step variables (PERLEN, NSTP, and TSMULT) in the Discretization input file (Harbaugh and others, 2000, p. 45). This may require defining more stress periods than would otherwise be required for the parent-grid model, child-grid model, or both.

Multiple Areas of Refinement

5. In the iterative coupling strategy discussed in this report, all child grids are simulated before re-simulating the parent grid with updated boundary conditions, as shown in figure 1 That is, the effects of all child grids are accumulated and applied to the parent grid in a single parent simulation. In this approach, the ordering of the child-grid simulations does not affect the convergence or the results. An alternative coupling strategy would be to iterate between a parent-child pair before simulating the next parent-child pair. In this approach, the ordering of the parent-child simulations could affect model performance. It is conceivable that such a strategy might be advantageous in certain circumstances (for example where hydraulic features in one grid have a much stronger influence on the flow system than the other grids); however, such an iterative coupling strategy was not investigated and is not currently (2013) available.

6. In the iterative coupling strategy used in this report, child model simulations can be executed independently of each other and, therefore, could be executed in parallel. However, parallelization of the child grid solutions is currently (2013) not implemented.

7. The execution time per iteration is approximately the execution time of the parent grid, T_{parent}, plus the sum of the execution times for each child, ΣT_{child}, which tends to cause a linear increase in execution time per iteration for each additional child model that is simulated. The number of LGR2 iterations required for convergence depends on the flow system being simulated. Experimentation so far has not indicated an increase in the number of LGR2 iterations when child grids are added.

Compatibility with Other MODFLOW Packages

LGR2 is integrated into MODFLOW-2005. It is designed to simulate the parent and child grids as two separate MODFLOW-2005 models and iterate between them until a balance of heads and fluxes along the interface between these models is achieved. Within each separate model, most MODFLOW-2005 packages can be used without alteration. Because the models are separate, different packages can be used for each model. For example, the parent model may use the DE4 solver while the child model uses PCG. This flexibility of iterative local grid refinement is one of its major advantages.

The ghost-node method does not use constant-head boundary cells (IBOUND <0) in the child model where it interfaces with the parent. Therefore, unlike the shared-node method, the input files of the individual models do not need to be changed to accommodate the interface cells when using the ghost-node method.

The parent grid uses specified-flux boundary conditions at the interface. If a constant-head boundary is specified in the parent model at the parent/child interface, the specified fluxes have no effect on the parent model. However, the influence of the constant head on the child model will still be present through the head interpolation of the ghost nodes. If the action of the specified-flux boundary condition is thought to be

important at this location, the constant-head boundary could be approximated by a general-head boundary with an appropriately large value of conductance.

Table 1 provides additional comments about the compatibility of ghost-node local grid refinement and MODFLOW-2005 Packages.

Running Parent and Child Models Independently Using the Boundary Flow and Head (BFH2) Package

The parent and child models can be simulated independently by using the coupling flux and head boundary conditions produced by LGR2. This can be accomplished using the new Boundary Flow and Head (BFH2) Package, which reads the coupling boundary conditions saved by LGR2 and replaces the original version of BFH in MODFLOW-LGR. Running the models independently can be useful when simulating solute transport, particle tracking, or other processes that do not affect the coupling boundary conditions produced by LGR2. This can result in substantial savings in computational time versus having to simulate the coupled parent-child system.

Situations that might affect the coupling boundary conditions, such as changes in pumpage, can also be simulated using an independent child or parent model, but simulated results lose accuracy because the coupling boundary conditions do not include the hydraulic effect of the changes. An analysis provided by the BFH2 Package can be used to determine if changes to either the parent or the child model require re-running LGR2 to update the coupling boundary conditions. For example, consider a situation in which after running LGR2 and finding coupling boundary conditions, the parent model is updated to include new pumping data for a well outside the refined area. How much does this well change heads and fluxes along the interfacing boundary where the child model

Table 1. Compatibility issues and possible adjustments required by the user for MODFLOW-2005 Groundwater Flow Process Packages.

Supported Packages[1]	Comments
BCF, LPF, HUF	Accuracy may be better if hydraulic properties are the same for adjoining cells at the interface. Different flow Packages can be used for the parent and child grids.
DIS	The time steps need to be identical in the coupled models. The grid refinement ratio needs to be integer: 1:1, 2:1, 3:1, 4:1, 5:1, and so on.
DE4, PCG, SIP, LMG, GMG, PCGN	Different solvers can be used for the parent and child grids.
SFR, STR, MNW, DRT	These Packages can be used within each grid but routing water across the interface between model grids currently is not supported in STR, MNW and DRT. Routing water across grids is supported in SFR.

[1]The three letter acronyms identify Groundwater Flow Process Packages. See the list of acronyms preceding the abstract of this report for definitions

is coupled? BFH2 Package output can be used to answer this question.

Instructions for the BFH2 package are presented in Appendix 2.

Using LGR2 to Simulate Solute Transport and Particle Tracking

Solute transport and particle tracking that are limited to the parent or child grid are simulated easily. Programs such as MT3DMS (Zheng and Wang, 1999) and MODPATH (Pollock, 1994) that act as post processors can use the binary cell-to-cell flow files produced by parent or child models for LGR2 simulations. For transport simulations that use the flow solution internally, such as the GWT Process, independent child or parent simulations can be run. In this case, first an LGR2 simulation is used to produce the coupling boundary conditions. Then transport is simulated with an individually run model that uses the BFH2 Package to impose the coupling boundary conditions.

Advective transport, simulated using particle tracking across grid boundaries, is supported by MODPATH-LGR (Dickinson and others, 2011) for the shared-node coupling method only. Currently (2013), MODPATH-LGR cannot be used with LGR2.

The simulation of transport by other processes such as dispersion and reactions across grid boundaries is not supported by any MODFLOW-compatible program as of this writing, but can be accomplished by using simulated results from the parent model as initial and boundary conditions for the child model. That is, this can be accomplished by simulating the models independently and employing one-way coupling of concentration initial and boundary conditions (Heywood, 2013). However, solute transport across interfacing grid boundaries is difficult to represent accurately because the abrupt change in grid size. For example, often the parent grid size is large enough that substantial numerical dispersion would be expected either when transport is simulated within its boundaries or if transport is simulated from a region where the grid is refined to a region where the grid is not refined. Ideally, the grids should be designed such that important features of any solute plume remain entirely within a single refined grid.

Description of Local Grid Refinement (LGR)

The function of the child model is to simulate phenomena that need a finer grid than the parent-model grid. For example, relatively fine grids are often needed to represent accurately sharp changes in hydraulic gradient, abrupt changes in hydraulic properties that would otherwise lose resolution if represented by the coarser parent grid, or other processes such as solute transport for which a fine grid is often needed to obtain

accurate solutions. The role of the parent model is to provide the boundary conditions to the child model that are consistent with the more regional flow system.

LGR2 uses the iteratively coupled ghost-node method of local grid refinement. Dickinson and others (2007) and Vilhelmsen and others (2011) indicate that the numerical properties of the ghost-node method are likely to be similar to those of the shared-node method, which was developed and tested extensively by Mehl and Hill (2005 and 2007). The basic procedure for solving the flow equation in a locally refined model is shown in figure 1 and the basic program flow of LGR2 within MODFLOW-2005 is shown in figure 2. The components of the iterative coupling are discussed in the following sections.

Grid of the Ghost-Node Method—Parent Grid, Child Grid, and the Interface

The grid structure is defined by how the sides, top, and bottom of the child model are nested within the parent model. The two models join along cell interfaces. The lateral boundaries of the child model may interface with the parent model or the sides of the simulated system. The top boundary of the child model is always the top of the simulated saturated groundwater system of the parent model. The bottom boundary of the child model may coincide with the bottom of the simulated system or may be an interface within the parent model.

Lateral Interface between the Parent and Child Grids

The lateral interface forms the sides of the child model grid. A schematic through the center of one layer of a locally refined grid is shown in figure 3. The interior cells of the parent model covered by the child model are made inactive by LGR2 by setting IBOUND to zero for these parent-model cells. Thus, after an initial parent-grid solution, the parent model has a hole in it that is filled by the child model. The parent and child models do not overlap cell areas. The model cells along the parent-child interface end at cell faces.

The ghost-node method described in this report requires a child-grid spacing that is an integer factor of the corresponding parent-grid spacing. For example, in the grid presented in figure 3, two child cells span the width of one parent cell, producing what is referred to as a 2:1 refinement ratio. Furthermore, the refinement along rows and columns needs to be the same in the two directions for all rows and columns. These are not requirements of the ghost-node method in general, rather, characteristics of this implementation.

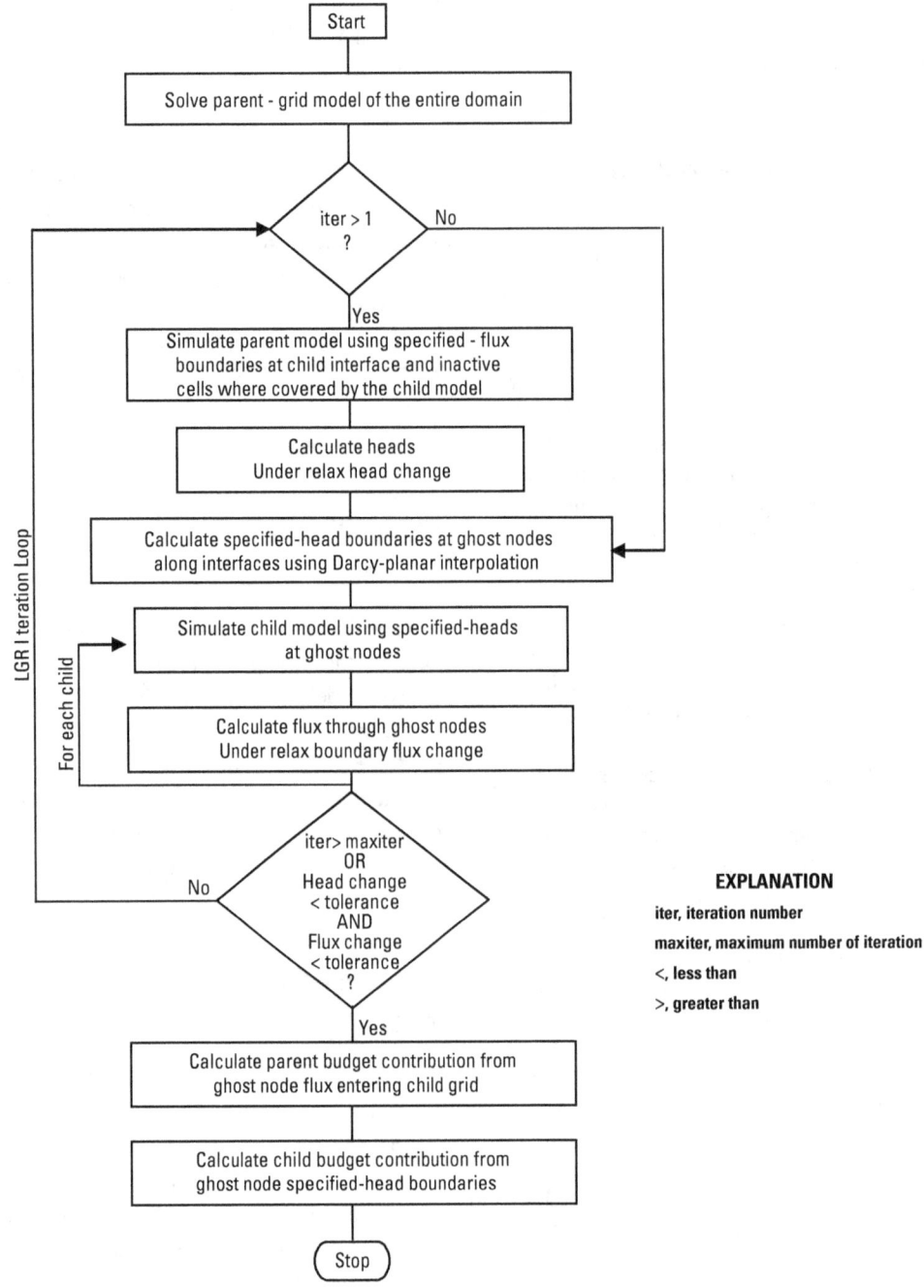

Figure 1.   Flow chart for the iteratively coupled Local Grid Refinement procedure with multiple child grids.

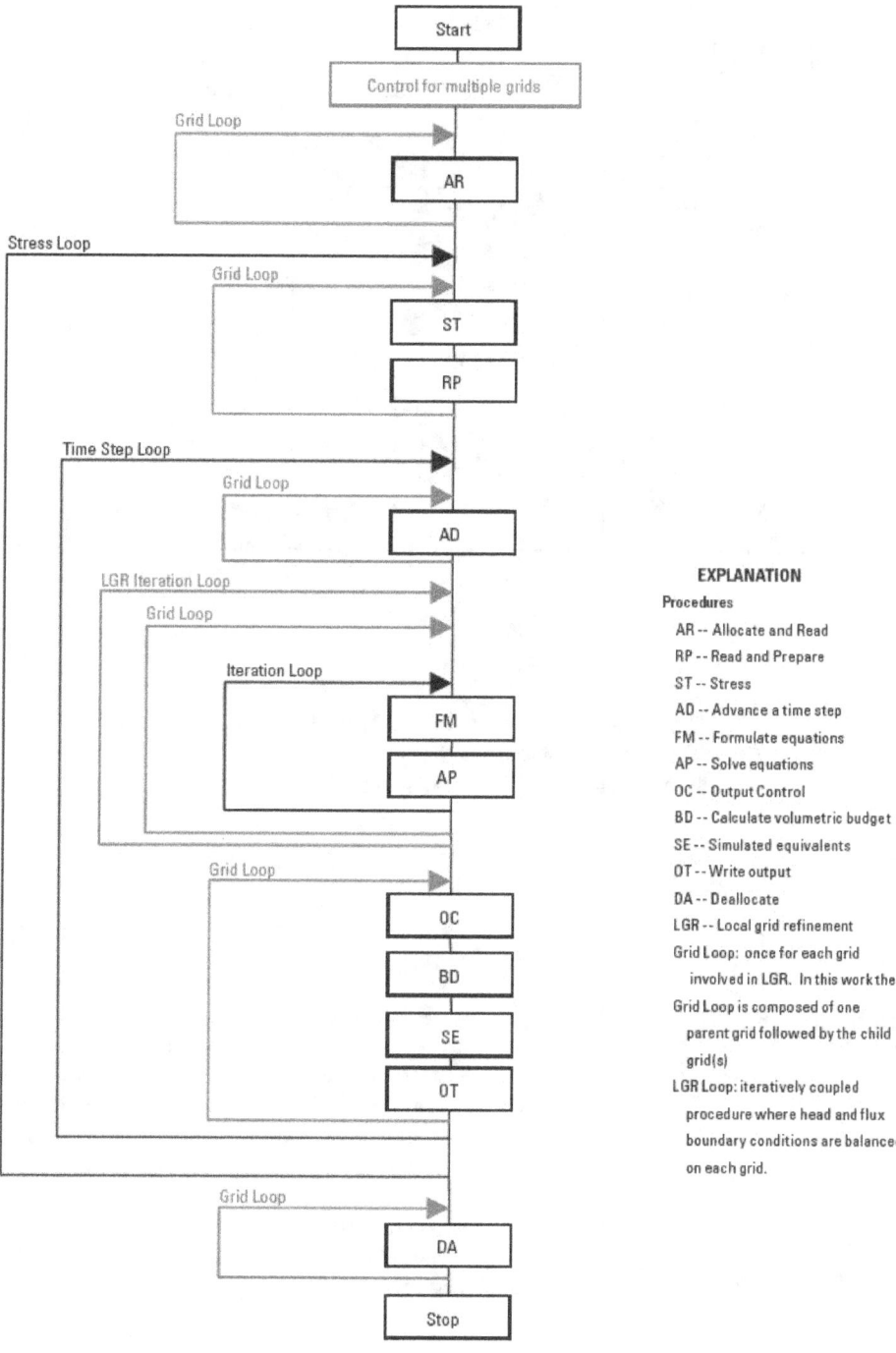

Figure 2.   Flowchart of MODFLOW-2005 with Local Grid Refinement.

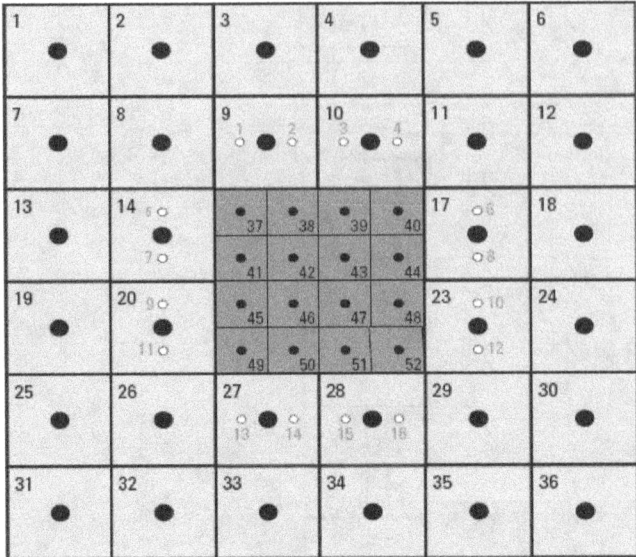

EXPLANATION

1   node number

○   ghost-node

1   ghost-node number

Figure 3. Two-dimensional areal schematic through the center of the locally refined grid. Cells of the child grid are darkly shaded; cells of the parent grid are lightly shaded.

The Top and Bottom of the Child Grid and Vertical Refinement

The top of the child model must coincide with the top of the simulated system of the parent model. The bottom of the child model can coincide with the bottom of the parent model or any cell face of any parent-model layer. Figure 4 shows the flexibility of vertical refinement possibilities.

For single-layer parent models (figure 4A), a vertically refined child model can be used. In this case, the heads at the ghost nodes at all elevations are set to the value obtained by interpolating the heads from the parent model along the boundary. This means no vertical gradients are simulated along the interface.

The uppermost ghost node(s) are between the top of the model and the uppermost nodes of the parent model grid. In this case, the ghost node(s) directly above the uppermost parent nodes are set to the value of head at the parent or associated interpolated node. This is consistent with a parent grid that has no vertical flow between the top of the model and the uppermost node. Vertical flows from recharge or discharge will not, therefore, be simulated correctly in the child grid at this location. In most circumstances, this is not expected to affect simulated results adversely.

Vertical refinement can vary layer by layer. For example, extra refinement at the top is illustrated in figure 4B where the top parent layer is refined vertically 5:1 and the second parent layer is refined vertically 2:1. Figure 4C shows that although the child refinement begins at the top layer, a 1:1 ratio can be used. Thicker upper layers may be helpful for problems with rewetting (see discussion in the "Unconfined Conditions" section).

If the child model extends to the bottom of the parent model (figure 4C), the ghost node(s) that are directly below the bottommost parent node are set to the value of head at the parent node or interpolated node above. This is consistent with a parent model that does not have vertical flux between the bottom of the system and the bottommost node.

For multi-layer parent models, single layer child models are not possible, as shown in figure 4D. This is because vertical hydraulic conductivity is not defined for single-layer models and thus connecting vertically to the underlying parent layer is not possible. If refinement is desired only in the uppermost parent layer, then a 2:1 vertical refinement in the child model would suffice.

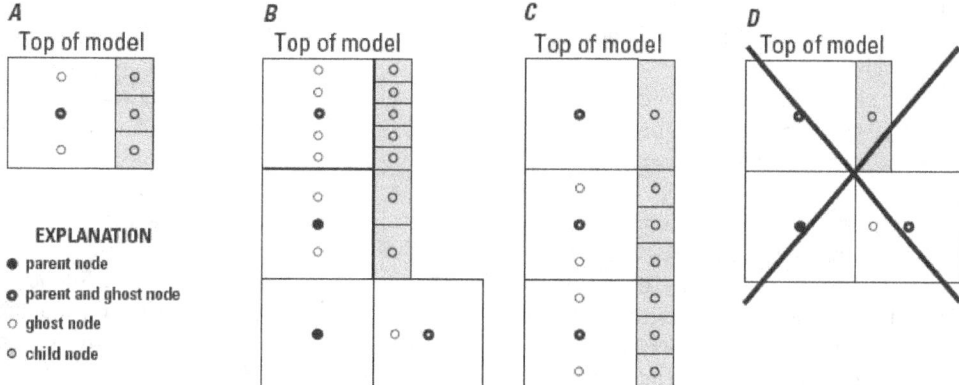

Figure 4. Cross-sectiional schematic of vertical refinement interface of *A*, a one-layer parent model refined to a three-layer child model; *B*, a multi-layer parent model where the child refinement varies vertically using both even and odd refinement ratios and terminates at the bottom of the second parent layer; *C*, a multi-layer parent model where the child refinement varies vertically using a 1:1 refinement ratio in the first layer and a 3:1 refinement ratio in subsequent layers. The refinement extends to the bottom of the parent model; and *D*, a multi-layer parent model with a single layer child model, which is not possible.

Including Multiple Areas of Refinement

The program documented in this report can simulate one or more block-shaped volumes of local-grid refinement. As currently implemented (2013), the maximum number of child grids is limited to nine. This number can be increased by increasing the allocation of derived type variables in each Fortran module to the total number of grids needed (see Harbaugh, 2005, p. 9-3–9-5). The refined areas cannot overlap, and at least two parent cells are needed between refined areas. Figure 5 shows an example with two parent-model grid cells between the two child model grids.

In three-dimensional models, different local grids can extend vertically to different levels of the parent grid. The vertical discretization needs to follow the same criteria described in the previous section.

The Iterative Coupling

The iteratively coupled method of LGR2 balances heads and fluxes across the interfacing boundary of the two grids. This is accomplished by iteratively updating the head (ghost nodes of child grid) and flux (parent grid) boundary conditions along the interface for each model. Depending on the formulation used, relaxing (averaging) with the head and flux values from the previous iteration is needed to keep the iterations stable (Funaro and others, 1988; Székely, 1998). This

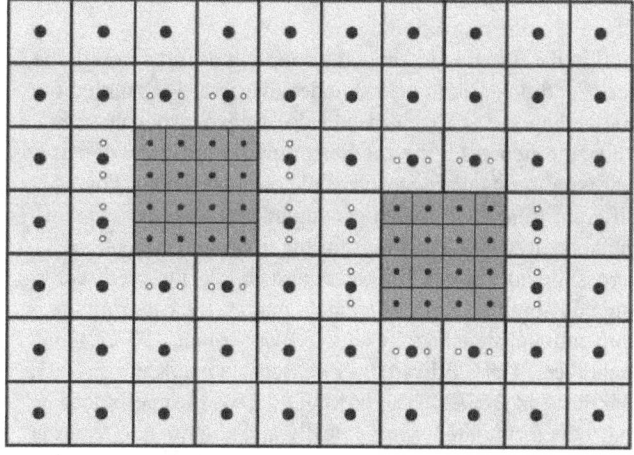

Figure 5. Schematic of a coarse grid with two areas of local grid refinement separated by two parent grid cells. At least two parent cells are required between child models.

approach of coupling the two grids is similar to what is used by domain decomposition methods (DDM). However, most DDM operate at the matrix level—they formulate the matrix equations first, and then break up the matrix equations into separate problems. LGR operates at the groundwater system level—the groundwater system is divided into parent and child grids, and then the equations for each are formulated. Similar approaches have been used by Funaro and others (1988) and Nacul (1991), which operate at the partial differential equation and reservoir level, respectively.

As shown in figure 1, the LGR procedure begins by simulating a parent model that encompasses the entire domain. For subsequent iterations, the parent-model cells completely covered by the child grid are eliminated. For the cells along the interface (figure 3), ghost nodes are located within the region of the parent model and linked to the child model. The heads from the parent model are used to interpolate the heads at the ghost nodes. The interpolated heads are relaxed using heads from the current and previous iteration (new and old heads, respectively, equation 1a). The value of the relaxation parameter may have to be adjusted to achieve convergence and is problem dependent. The child model is simulated and the fluxes through the ghost nodes at the parent/child interface are calculated and also relaxed (equation 1b) before being used as the parent flux boundary condition. The parent model is simulated using these updated flux boundary conditions and produces updated heads for the interpolation onto the ghost nodes that are linked to the child model boundary. This process is repeated until both the head change and the flux change at the ghost nodes are smaller than user-defined criteria.

$$\text{head}^{\text{updated}} = \omega \cdot \text{head}^{\text{new}} + (1-\omega) \cdot \text{head}^{\text{old}} \tag{1a}$$

$$\text{flux}^{\text{updated}} = \omega \cdot \text{flux}^{\text{new}} + (1-\omega) \cdot \text{flux}^{\text{old}} \tag{1b}$$

where ω is the relaxation factor.

In the iterative method, the coupling occurs through the heads and fluxes at the ghost nodes, which are accounted for on the diagonal and right-hand side of the matrix equations. Thus, the stencil for the coefficient matrix is always consistent with the standard stencil of the original model. This is different from other two-way coupled local grid-refinement methods, in which equations for the irregular connections across the interface of the parent and child grids are directly embedded into a single coefficient matrix, thus altering the conventional stencil (for example, Wasserman, 1987; Ewing and others, 1991; Edwards, 1999; Schaars and Kamps, 2001; Haefner and Boy, 2003). For MODFLOW, the coefficient matrix is formulated symmetrically and all non-zero terms are located on the matrix diagonal and six off diagonals (McDonald and Harbaugh, 1988, p. 12-2–12-4). Therefore, when using the iterative coupling considered in this work, efficient solvers that are based on a conventional finite-difference stencil on a Cartesian grid, such as the solvers distributed with MODFLOW, can be applied without modification. The resulting matrix equations are:

$$[A_p]\{h_p\} = \{f_p(h_c)\} \tag{2a}$$
$$[A_c]\{h_c\} = \{f_c(h_p)\} \tag{2b}$$

where,

$[\ ]$ denotes a matrix and $\{\ \}$ denotes a vector,

$[A_p]$ and

$[A_c]$ are the coefficient matrices for the parent and child grid, respectively. They contain conductances and storage properties and have the same structure as a conventional finite-difference discretization,

$\{h_p\}$ is the head in the parent grid,

$\{h_c\}$ is the head in the child grid,

$\{f_p(h_c)\}$ is the right-hand side for the parent grid, and includes the flux boundary condition through the ghost nodes. This flux is determined from ghost-node conductance and the difference between the head at the ghost node and heads in the child grid from the previous iteration (h_c). It also contains the storage terms from the previous time step and all other boundary conditions and stresses within the parent model, and

$\{f_c(h_p)\}$ is the right-hand side for the child grid and includes the ghost node specified-head boundary condition along the interface with the parent grid. This boundary condition is determined from the previous head solution on the parent grid (h_p) using interpolation. It also contains the storage terms from the previous time step and all other boundary conditions and stresses within the child model.

A common approach to handling asymmetric matrices is to iteratively solve them by using symmetric solvers and splitting the coefficient matrix such that the asymmetric terms are evaluated on the right-hand side of the matrix equations. For example, the vertical flow calculation under dewatered conditions (McDonald and Harbaugh, 1988, p. 5-21–5-23) uses this type of splitting as does the Layer Variable-Direction Horizontal Anisotropy (LVDA) capability of the Hydrologic-Unit Flow (HUF) Package (Anderman and others, 2002). In this regard, the iterative method outlined above can be viewed as a matrix splitting of a directly embedded approach, where the coupling terms that are not involved in the conventional stencil are placed on the right-hand side of the matrix equations. Head values from the previous iteration are used to evaluate these terms.

The details of how the ghost node coupling equations and boundary conditions are calculated are discussed in the following sections.

Ghost-Node Coupling

The ghost nodes provide the linkage between the two grids. The child model is simulated with the ghost-nodes used as a specified head boundary condition with appropriate branch conductance (analogous to a General Head Boundary in MODFLOW). The parent model is simulated with specified flux boundary conditions that are calculated from the product of the branch conductance and the head difference between the ghost node and the adjacent child head.

In figure 3, consider the mass balance equation for parent cell 9 in terms of volumetric flow (Q) to/from adjacent cells:

$$Q_{8-9} - Q_{9-10} + Q_{3-9} - Q_{9-37} - Q_{9-38} = 0 \qquad (3)$$

Writing the flows in terms of conductances between cells and adjacent heads:

$$CR_8 \times (h_8 - h_9) - CR_9 \times (h_9 - h_{10}) + CC_3 \times (h_3 - h_9) - CGN_1 \times \atop (h_{GN1} - h_{37}) - CGN_2 \times (h_{GN2} - h_{38}) = 0 \qquad (4)$$

Collecting terms results in:

$$-[CR_8 + CR_9 + CC_3] \times h_9 + CR_8 \times h_8 + CR_9 \times h_{10} + CC_3 \times h_3 \atop = CGN_1 \times (h_{GN1} - h_{37}) + CGN_2 \times (h_{GN2} - h_{38}) \qquad (5)$$

Consider the mass balance equation for child cell 37 in terms of volumetric flow to/from adjacent cells:

$$Q_{14-37} - Q_{37-38} + Q_{9-37} - Q_{37-41} = 0 \qquad (6)$$

Writing the flows in terms of conductances between cells and adjacent heads:

$$CGN_5 \times (h_{GN5} - h_{37}) - CR_{37} \times (h_{37} - h_{38}) + CGN_1 \times (h_{GN1} - h_{37}) - \atop CC_{37} \times (h_{37} - h_{41}) = 0 \qquad (7)$$

Collecting terms results in:

$$-[CGN_5 + CR_{37} + CGN_1 + CC_{37}] \times h_{37} + CR_{37} \times h_{38} + \atop CC_{37} \times h_{41} = -CGN_5 \times h_{GN5} - CGN_1 \times h_{GN1} \qquad (8)$$

where

CGN_x is the branch conductance between a ghost node and the adjoining child node, and referred to as the ghost node conductance in this report. Similar to other branch conductances, CC and CR, the ghost node conductance is based on the equivalent conductance of two prisms in series (McDonald and Harbaugh, 1988, equation. 39) for flow between the ghost node and the adjoining child node, as shown in figure 6 for 2:1 refinement both horizontally and vertically. Using the values indicated in figure 6, the ghost node conductance is:

$$CGN = [COND_p \times COND_C]/[COND_p + \atop COND_C] \qquad (9a)$$

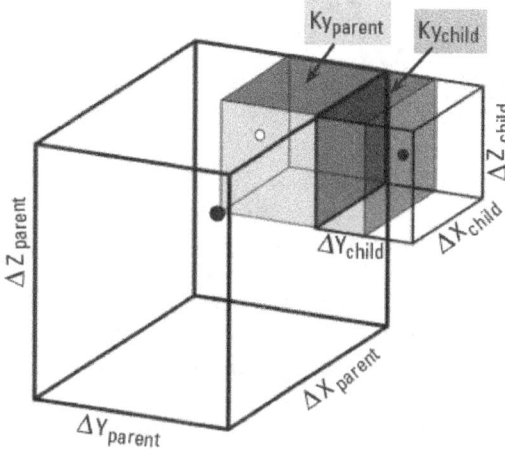

Figure 6. Schematic of a 2:1 vertically and horizontally refined grid interface showing shaded prisms between the ghost node and adjacent child node that are used in the calculation of ghost node conductances. $K_{Yparent}$ and K_{Ychild} are the hydraulic conductivities in the y direction for the parent and child grids, respectively.

$$COND_p = [Ky_{parent} \times \Delta x_{child} \times \Delta z_{child}]/[\Delta y_{parent}/2] \qquad (9b)$$
$$COND_C = [Ky_{child} \times \Delta x_{child} \times \Delta z_{child}]/[\Delta y_{child}/2] \qquad (9c)$$

where

K_x, K_y, and K_z and Δx, Δy, and Δz are the hydraulic conductivities and the cell spacing in the x, y, and z,directions, respectively, for either the parent or child grid. Similar equations are written for the other cells along the grid interface.

The heads at the ghost-nodes are solved using a Darcy-planar interpolation based on the heads at the adjacent parent cells. This is discussed in the "Determining Ghost Node Specified Heads" section.

Two Methods of Formulating and Solving the Equations

Two methods are considered for formulating and solving the ghost-node coupled equations; both methods require iterations between grids to converge to a final solution.

The first method uses equation 5 for the parent interface cells and equation 8 for the child interface cells. In this case, the right-hand side (RHS) of equation 5 corresponds to the flux between the ghost node and the adjoining child cell. This approach (referred to as "method 1") is analogous to the formulation used in the shared-node method of LGR (Mehl and Hill, 2005), and also requires under relaxation of heads and fluxes for a stable solution.

The second method (referred to as "method 2") rewrites the ghost-node head (h_{GN}) as the head (h) at the parent cell plus some change in head (Δh_{GN}). For example, for parent cell 9:

$$h_{GN1} = h_9 + \Delta h_{GN1} \qquad (10)$$

Substituting equation 10 into equation 5 and collecting terms results in:

$$-[CR_8 + CR_9 + CC_3 + CGN_1 + CGN_2] \times h_9 + CR_8 \times h_8$$
$$+ CR_9 \times h_{10} + CC_3 \times h_3 = CGN_1 \times (\Delta h_{GN1} - h_{37}) + \qquad (11)$$
$$CGN_2 \times (\Delta h_{GN2} - h_{38})$$

For method 2, the child interface cells use the same formulation as method 1 given in equation 8.

Method 2 does not require under relaxation of heads for a stable solution. This is probably because method 2 can be considered more implicit than method 1, because the parent head and conductance terms associated with ghost nodes appear in the left-hand side of the equations rather than applying the flux from the previous child solution to the RHS. When using method 2, it was found that over relaxation of the ghost-node heads improved convergence. How to apply relaxation to the terms associated with the fluxes is not straightforward in this formulation. Limited testing has shown that method 1 requires fewer iterations to achieve convergence, but method 2 is retained as a user option as there might be cases where the stability of this method is advantageous.

Method 2 can also be viewed as a flux splitting technique that partitions the flux between a parent cell and the adjoining child cells into how much flux is due to the parent head and how much is due to the variation in head represented by the ghost nodes. Similar flux splitting techniques are used by the LVDA capability of the HUF Package (Anderman and others, 2002).

Determining Ghost Node Specified Heads

For ghost nodes that are coincident with the parent model nodes (fig. 7, dark circles with white centers), the heads calculated by the parent model apply directly. For ghost nodes that are not coincident with a parent node (open circles with a dashed outline), the head is interpolated. Linear or other low-order, geometrically based polynomial interpolation has been suggested (Quandalle and Besset, 1985; Ward and others, 1987; Ewing and others, 1991; Leake and others, 1998; Székely, 1998; Davison and Lerner, 2000). Figure 7 shows that in the presence of heterogeneity, linear interpolation produces heads that do not obey the physics of groundwater flow. Other geometric interpolation methods share this difficulty. For this reason, an alternative Darcy-planar interpolation method is used in LGR2 that circumvents this problem. The interpolation is described in the following paragraphs. First, the concepts are illustrated analytically using a one-dimensional interface boundary of a two-dimensional model. The implementation used by LGR2 is then described.

Interpolation Concepts Illustrated Analytically Using a Two-Dimensional Model

The fundamental constitutive relation that governs heads and fluxes in groundwater systems is Darcy's law. In one dimension:

$$q = -K(dh/dx) \qquad (12)$$

where

q is the flux (Flow rate per unit area or Darcy velocity),

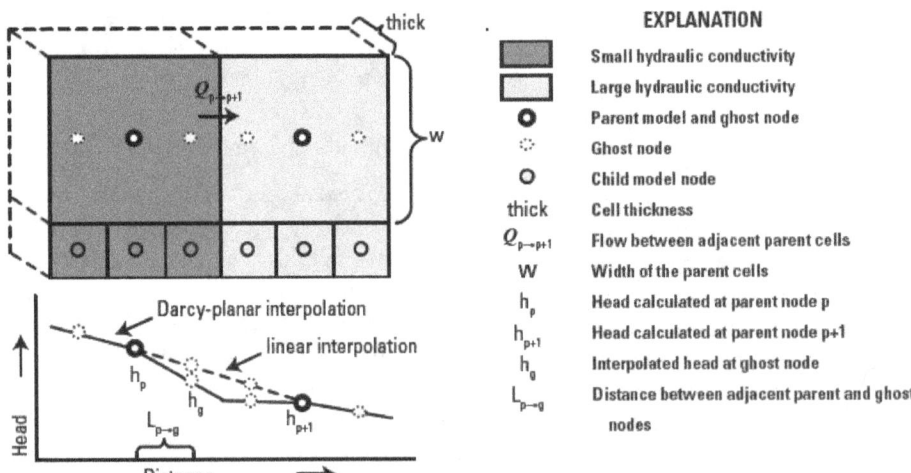

Figure 7. Darcy-planar interpolation in relation to linear interpolation between ghost nodes for cells with different hydraulic conductivity, as denoted by the different shading of the two cells.

K   is the hydraulic conductivity, and

dh/dx   is the hydraulic gradient.

Darcy's law implies that if the one-dimensional flux, the material properties, and the physical dimensions are all known between two points, the hydraulic gradient (head loss) at each location between the two points can be uniquely determined. Interpolating the heads along the boundary of the child grid using Darcy's law produces heads that are consistent with the parent-grid flow field. This interpolation also is consistent with finite-difference discretization of the groundwater flow equations. The resulting interpolation scheme, referred to here as Darcy-weighted interpolation, is calculated as:

$$h_{GN} = h_p - \left(\frac{Q_{p \to p+1}}{K_p} \times \frac{L_{p \to g}}{A_p} \right) \qquad (13)$$

where

h_{GN} is the head at the ghost node,

h_p is the head at the parent node,

$Q_{p \to p+1}$ is the flow between adjacent parent cells

K_p is the hydraulic conductivity of the parent cell,

A_p is the cross-sectional area of the parent cell perpendicular to flow,

$L_{p \to g}$ is the distance between parent node and ghost node.

Various forms of one-dimensional Darcy-based interpolation have been developed and used by others (Wasserman, 1987; Schaars and Kamps, 2001; Haefner and Boy, 2003). For confined flow, they all produce the same interpolated heads, and these heads are consistent with the flow of the parent grid. However, modifications are needed to extend them to the two-dimensional interfaces of three-dimensional models. A one-step method referred to here as "Darcy-planar interpolation" was developed to interpolate heads at ghost nodes for two-and-three dimensional models.

Darcy-Planar Interpolation

The method described above can be used to form the basis for extension to three dimensions. The term in parenthesis in equation 13 calculates the head change from head loss in a given flow direction. For interpolating heads onto the ghost nodes on a two-dimensional plane (as is needed for three-dimensional models), head losses in the other flow directions are needed. Extending equation 13 to include the other flow directions results in:

$$\begin{aligned} h_g = h_p &- \left(\frac{Qx_{p \to p+1}}{Kx_p} \times \frac{Lx_{p \to g}}{Ax_p} \right) \\ &- \left(\frac{Qy_{p \to p+1}}{Ky_p} \times \frac{Ly_{p \to g}}{Ay_p} \right) \\ &- \left(\frac{Qz_{p \to p+1}}{Kz_p} \times \frac{Lz_{p \to g}}{Az_p} \right) \end{aligned} \qquad (14)$$

where the variables are the same as defined in equation 13, except that the addition of x, y, and z indicate the components in those respective coordinate directions. Equation 14 is applicable to both two-and-three dimensional models. When using the BCF or HUF Packages, vertical hydraulic conductivity is formulated between nodes rather than for a cell, which results in a vertical linear interpolation. Because the ghost nodes are aligned with the parent nodes in at least one coordinate direction, at least one of the three terms in parenthesis in equation 14 will be zero. For ghost nodes that are at the edge of the parent model, the component of flow perpendicular to the boundary is assumed to be zero through the model boundary, and the head loss in that direction is zero. Therefore, the ghost-node head is the same as the head at the parent node for ghost nodes in the corner of a parent model.

This implementation produces the same interpolated heads as the cage-shell interpolation used by the shared-node method (Mehl and Hill, 2005) for steady-state, confined simulations of three-dimensional homogeneous or two-dimensional heterogeneous systems. It is similar to the Darcy-weighted interpolation described by Dickenson and others (2007). However, unlike the method used by Dickenson and others, the Darcy-planar method can be formulated in a single step rather than having to interpolate sequentially in each direction. The accuracy of the Darcy-weighted method used by Dickenson and others depends on which direction is interpolated first, because each direction can have a different accuracy. Limited comparisons of these methods showed that the Darcy-planar interpolation was always more accurate than the least accurate direction of the Darcy-weighted method. For this reason, and because it can be formulated in a single step (which makes it easier to include in a formulation that solves all grids in a single coefficient matrix), the Darcy-planar method is used in LGR2.

Unconfined Conditions

A disadvantage of the Darcy-planar interpolation used in LGR2 compared to the cage-shell interpolation used in the shared-node method is that interpolated heads are not as accurate in unconfined conditions. This is because the cage-shell interpolation solves for the interpolated heads numerically, which accounts for the nonlinearity of unconfined flow, whereas the Darcy-planar interpolation (equation 14) does not. However, if the rewetting capability is used and cells go dry along the parent/child interface, the cage-shell interpolation scheme no longer produces meaningful values.

An advantage of using the Darcy-planar interpolation is that realistic interpolated heads can still be produced when dry cells occur along the parent/child interface.

Although LGR2 can simulate conditions where dry cells occur along the parent/child interface, the drying and rewetting of these cells can exacerbate convergence problems. For this reason, one might want to use thicker layers at the top of the model. While the child model must start at the top layer of the parent, grid refinement does not need to start at the top because a vertical refinement ratio of 1:1 can be used in any parent layer (see fig. 4C). Using thicker upper layers in the child model can be useful in alleviating some of the drying and rewetting problems associated with thin layers at the top of the model.

Closure Criteria

Closure criteria are needed to determine when to stop the LGR iterative procedure (see fig. 1) in addition to any closure criteria needed by the solver package. Closure criteria of the LGR iterations control the accuracy of both the head and flux boundary conditions, and thus control the quality of the overall LGR solution. Given these boundary conditions, closure criteria for the solver package used by MODFLOW-2005 control the accuracy of the parent and child solutions. Unlike standard MODFLOW-2005, MODFLOW-LGR will not terminate the simulation if either the parent or child model simulation did not converge to the closure criteria specified by their respective solver package.

Closure Criteria for LGR Iterations

Separate closure criteria are needed for the parent and child grids. For convergence of the parent grid, the maximum relative change of the coupling specified-flux boundary condition between successive iterations needs to be less than a user-defined amount (see equation 15a). For convergence of the child grid, the maximum change of the coupling specified-head boundary condition between successive iterations needs to be less than a user-defined amount (see equation 15b).

$$|flux^{i+1} - flux^i| / max(|flux^i|, 1.0) \qquad (15a)$$

$$|head^{i+1} - head^i| \qquad (15b)$$

where

flux	is the volumetric flux,
max( )	is the maximum of the given arguments,
superscript 'i'	indicates the LGR iteration, and
|·|	indicates the absolute value.

After convergence of the LGR iterations, the ghost-node flux across the parent/child interface of each child model should be examined. It is printed after the volumetric budget

in the listing file for each child model. The mass-balance calculation is based on the ghost-node fluxes along the parent/child interface. It is an indicator of the overall quality of the LGR solution because it shows how precisely the flow in/out of the parent grid through the ghost-nodes matches the ghost-node flows calculated by the subsequent child solution. If the mass balance is deemed too large, lower the closure criteria of the LGR iterations until an acceptable mass balance is achieved. More LGR iterations may be needed to achieve the lower closure criteria. Generally, the same guidelines that are often used for standard MODFLOW simulations can be applied here, so that differences in ghost-node flows of less than 1 percent are adequate. Because of errors introduced from the abrupt change in grid size, some mass-balance error may remain which cannot be attenuated with further iterations. Therefore, the LGR iterations also are stopped after the user-specified maximum number of LGR iterations is exceeded.

Conversely, if the LGR iterations do not converge, but differences between the parent and child ghost-node flows are small, then the quality of the LGR solution is probably acceptable. In this case, the closure criteria generally can be increased such that convergence is achieved and the mass balance at the parent/child interface is still acceptable. An example is shown in Appendix 1.

Solver Iterations

The available solvers for MODFLOW-2005—SIP, (McDonald and Harbaugh, 1988), PCG2 (Hill, 1990), DE4 (Harbaugh, 1995), LMG (Mehl and Hill, 2001), GMG (Wilson and Naff, 2004), and PCGN (Naff and Banta, 2008)—are compatible with LGR2. The parent and child models can use different solvers. Generally, the closure criteria used for the solvers should be less than or equal to what is used for the LGR closure. For example, it does not make sense to try to solve the coupling boundary conditions to a precision of 10^{-5} when the overall head solution, as controlled by the solver closure criteria, is only accurate to 10^{-3}. HCLOSE, which is head closure criteria in all solvers except LMG, can be compared to the head closure for the LGR iteration. Only rough guidelines can be provided here, but generally it is better to be cautious and use strict closure criteria for the solver. If this results in excessive computer processing times, adjustments to the solver closure criteria should be made.

Transient Simulations

For transient simulations, the iterative process described is repeated for each time step of each stress period, as shown in the flow chart of figure 2. As currently (2013) implemented, LGR2 requires that the parent and child grids use equivalent time discretization.

For transient flow, the Darcy-planar interpolation procedure does not maintain perfect consistency with the flow of the parent grid, even for simple one-dimensional transient flows.

This is because storage effects are not accounted for in the Darcy-planar interpolation (see equation 14) and transient flow phenomena are propagated differently through different grid sizes, as discussed by Mehl and Hill (2005, p. 21).

This error along the boundary has implications for volumetric budget calculations for large-scale regional models that are used as the parent grid. In such models, small changes in head due to coupling errors at the interface can result in large changes in storage. For example, consider a regional model with cell dimensions of 1000×1000 meters (m), a specific yield of 0.2, and coupled to a local model using LGR2 with a head closure criterion set at 1.0×10^{-2} m. Heads along the boundary can have errors on the order of 1.0×10^{-3}. This error, when viewed as a head change, can cause changes in storage on the order of $1.0\times10^{-3}\times1000\times1000\times0.2 = 200$ cubic meters (m^3), which may be a significant amount of the overall budget depending on the flow system. Changes in storage are calculated on a per time-step basis, as

$$\Delta S/\Delta t \tag{16}$$

where

ΔS	is the change in storage $= (h^{n+1}-h^n)\times S_y \times A_c$
Δt	is the time-step size
h^n	is head at time step n
S_y	is specific yield
A_c	is planar area of the cell

In accordance with equation 16, this error in storage is increased by smaller time steps and attenuated with larger time steps. Furthermore, within a given stress period, changes in storage decrease as time progresses, which reduces this error. Therefore, storage changes that occur at early time steps within a stress period can contain a significant amount of error, yet have reliable accuracy at later time steps. Users should keep these issues in mind when examining changes in storage for large-scale regional models coupled to local-scale models using LGR.

Examples

LGR2 has been tested under a variety of conditions to evaluate its convergence properties and numerical accuracy. The three examples presented in this section all involve synthetic test cases and some evaluation of the error. These test cases were described and used in the documentation of the shared node LGR (Mehl and Hill, 2005; 2007). Despite using a coupling that conserves mass between the two grids, an error is still introduced on the interface between the two grids because of the abrupt change in grid resolution. Therefore, it is important to evaluate the accuracy of this method. In this section both shared-node and ghost-node LGR are considered. Lastly, computational processing time is typically of concern when simulating groundwater models, so these quantities are

compared. These comparisons can be used to help modelers decide how to use LGR2 most advantageously.

Example 1: Two-Dimensional Steady State Test Case with Heterogeneity and Pumping

The accuracy of the grid-refinement technique presented in this work is compared with the shared node method of LGR for simulating flow in the two-dimensional, heterogeneous, confined aquifer with a pumping well shown in figure 8. The heterogeneity pattern is based on a laboratory experiment described by Garcia (1995) and Mapa and others (1994) and is described in the LGR documentation for the shared-node method (Mehl and Hill, 2005, p. 30). As shown in figure 8, the system has constant-head boundaries on the left and right side of 10.0 m and 1.0 m, respectively, and no-flow boundaries along the top and bottom. A pumping well extracts 5.5×10^{-3} cubic meters per second (m^3/s) from the system. Figure 9 shows the flow vectors and contours for this system when using the low-contrast set of transmissivity values listed in figure 8. Grid refinement is applied to increase the accuracy of the model in the vicinity of the well. Although this system is synthetic, and therefore limited in its applicability to real aquifer systems, the results are likely to provide insights regarding the typical performance that can be expected from applications of different methods of local grid refinement. The advantage of this test case is that the hydraulic-conductivity distribution provides a numerically difficult challenge for testing.

The accuracy of the techniques are evaluated by comparing the results from a uniform fine grid ("true" solution) to (a) the shared-node and (b) the ghost-node method of local grid refinement using the two different methods of coupling. All grids were constructed such that the finite-difference cells are always fully within a single hydraulic-conductivity block. The globally refined grid has 450 rows and 972 columns, with cell dimensions of 1.028 m and a 1.0 m in the horizontal and vertical directions, respectively. The parent grid has 50 rows and 108 columns, with cell dimensions of 9.25 m and 9.0 m in the horizontal and vertical directions, respectively. Although the same interior nodal locations are compared, because of the differences in how the grids connect, the ghost-node and the shared-node grids do not occupy the same area of the parent. Therefore, rigorous comparison of the results is not possible. Nevertheless, efforts were made to make the comparisons as close as possible. The ghost-node grid was given more rows than the shared-node grid (108 versus 100) and fewer columns than the shared-node grid (144 versus 154). This produces approximately the same total number of refined cells in the ghost-node grid as the shared-node grid (15,552 versus 15,400).

Calculation of Heads

The results of the comparisons are shown in table 2. All comparisons are made relative to the fine-grid model, such that the mean head error shown in the second column is calculated

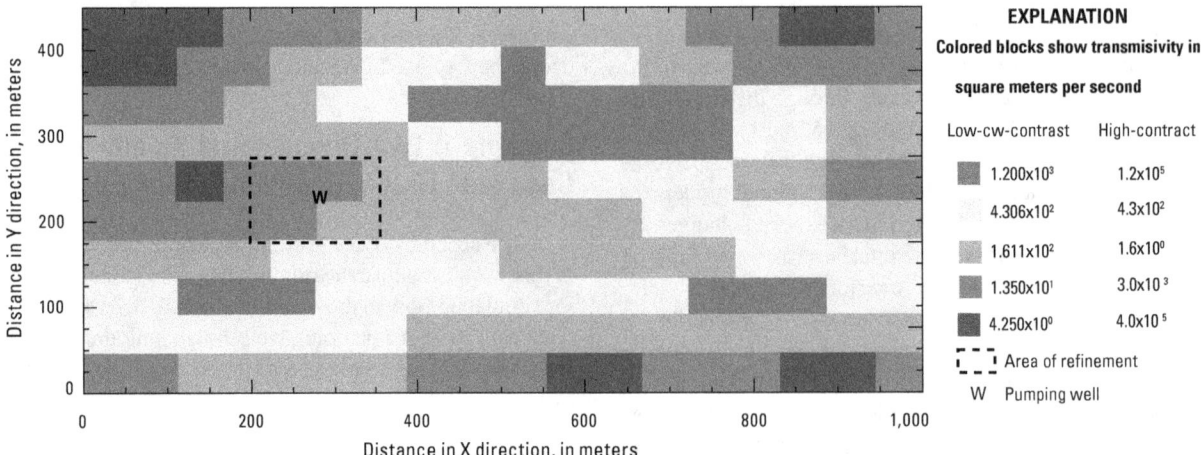

Figure 8. Heterogeneity structure and area of local refinement around the well indicated by dashed rectangle. Model results using the two sets of transmissivities are shown in figure 9 and figure 10, respectively

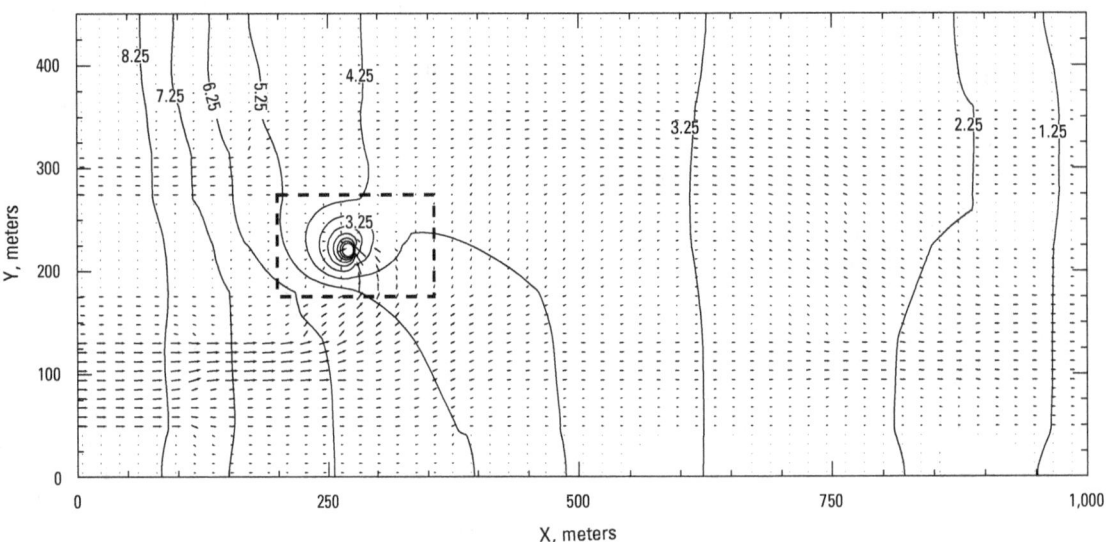

Figure 9. Flow vectors and head contours calculated using the low-contrast set of transmissivities listed in figure 8. Area within the dashed rectangle indicates region of local refinement.

Table 2. Comparison of errors and computer processing time for the shared-node and ghost-node grid refinement methods applied to the low-contrast version of example 1.

[The system is depicted in figure 8 with the low-contrast set of transmissivities that range approximately 2.5 orders of magnitude from 1 2x10³ to 4.25x10⁰ square meters per second. Computation times using a Linux workstation, Intel XEON—2.0 gigahertz (GHz), 2 gigabytes (Gb) Ram); %, percent; s, second; LGR, local grid refinement]

Gridding method	Mean interior head error (%)	Number of LGR iterations	Computer processing time (s)
LGR-shared node[1]	0.097	10	2.7
Ghost-node method 1[1]	0.307	9	2.7
Ghost-node method 2[2]	0.332	45	7.9

[1]Relaxation factor for heads and fluxes is 0.5.

[2]Relaxation factor for heads is 1.0.

as the L_1 norm of the differences between the model approximation and fine grid ("true" solution) normalized by dividing by the fine grid ("true" solution). All results were obtained using a preconditioned conjugate gradient solver, PCG2 (Hill, 1990) with both HCLOSE and RCLOSE set to 1×10^{-8}.

Effect of Heterogeneity Contrast

The degree of heterogeneity can have a substantial effect on the flow system, and consequently, on the efficiency and accuracy of the local grid-refinement methods. These effects are investigated by using the same system as shown in figure 8, except the magnitude of the hydraulic-conductivity field is changed such that the contrasts between materials is increased. The new values of transmissivity are shown as the

high-contrast set of values in figure 8, and the flow vectors and head contours for this system are shown in figure 10. Comparison of figure 9 and figure 10 shows that the increase in heterogeneity contrasts causes the hydraulic gradients through the system to be much steeper in some locations and change direction very rapidly, particularly in the region of local refinement.

The same methods of grid refinement were evaluated and the same solvers and settings were used for this hydraulically more complicated system. Results are summarized in table 3. A surprising result is that the ghost-node simulation using method 2 requires fewer iterations than when simulating the less heterogeneous system (19 versus 45). It also has the least amount of error.

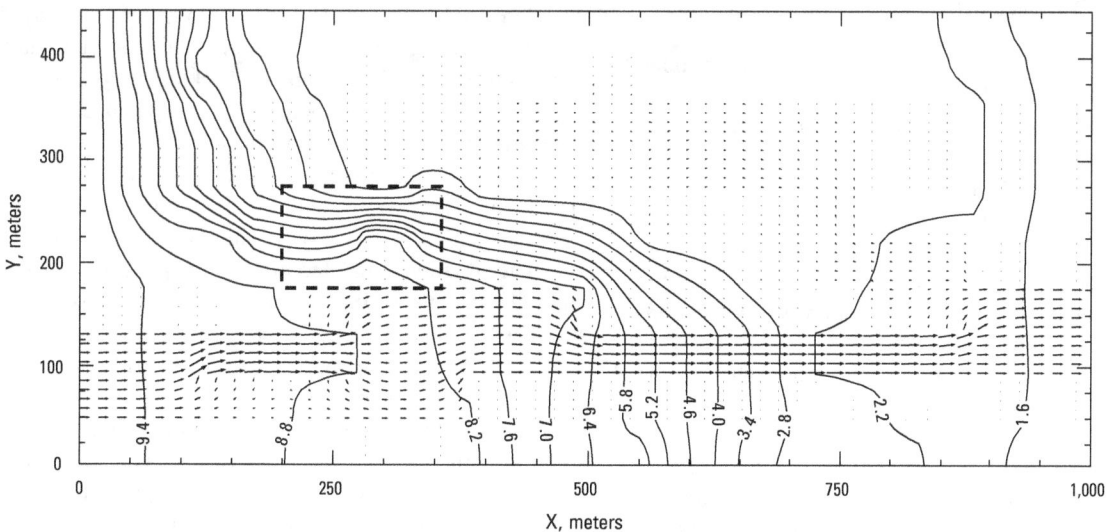

Figure 10. Flow vectors and head contours calculated using the high-contrast set of transmissivities listed in figure 8. Area within the dashed rectangle indicates region of local refinement.

Table 3. Comparison of errors and computer processing time for the shared-node and ghost-node grid refinement methods applied to the high-contrast version of example 1.

[The system is depicted in figure 8 with the high-contrast set of transmissivities that range approximately 10 orders of magnitude from 1.2×10^5 to 4.0×10^{-5} square meters per second. Computation times using a Linux workstation, Intel XEON—2.0 gigahertz (GHz), 2 gigabytes (Gb) Ram); %, percent; s, second; LGR, local grid refinement]

Gridding method	Mean interior head error (%)	Number of LGR iterations	Computer processing time (s)
LGR-shared node[1]	0.079	14	4.8
Ghost-node method 1[1]	0.063	9	3.6
Ghost-node method 2[2]	0.059	19	5.4

[1]Relaxation factor for heads and fluxes is 0.5.

[2]Relaxation factor for heads is 1.0.

These comparisons show that the ghost-node coupling (with either method 1 or method 2) produces results that have similar accuracy as the shared-node coupling. From these results, it is not clear that one method is more accurate than the other. Also, the number of coupling iterations and required CPU times are fairly similar for the shared-node coupling and the ghost-node coupling using method 1. In these tests, and other tests not reported here, method 1 always required fewer iterations and less CPU time than method 2. The final results of method 1 and method 2 are not identical because they converge differently; this results in slightly different values at the converged solution. As the LGR closure criteria are decreased, these differences in the converged solution decrease.

Example 2: Three-Dimensional, Unconfined, Transient Test Case with Heterogeneity, Pumping, and Rewetting

This test case illustrates the capability of the ghost-node method to simulate unconfined conditions where wetting and drying occur. The parent model has 19 rows and columns and 3 layers. The cell dimensions of 10 meters are uniform along all rows and columns. The layers have equal thickness of 16.667 m; the top layer is unconfined with a top elevation of 50 m. The child model has 15 rows and columns and 5 layers. It is a 3:1 horizontal refinement of the parent grid resulting in uniform cell dimensions of 3.333 m for all rows and columns. The horizontal refinement begins in row 8, column 8, and ends in row 12, column 12 of the parent model. The vertical refinement extends over the upper two layers of the parent model. The vertical refinement ratio varies by layer with a 1:1 ratio in the first parent layer and a 4:1 vertical refinement ratio for the second parent layer. A pumping well located in the center of the aquifer extracts a total of 900 m^3 per day and is simulated as 300 m^3 per day from layers 2, 3, and 4 of the child model. Constant-head boundaries of 50 m and 41 m are located along the west and east edges of the domain, respectively. All other boundaries are no flow.

The aquifer is heterogeneous, but homogenous within a layer. The top layer is unconfined and has horizontal hydraulic conductivity, vertical hydraulic conductivity, specific yield, and specific storage of 1.0 meters per day (m/day), 1.1 m/day, 0.2, and 5×10^{-5} per meter, respectively. All other layers are confined and have the same specific storage as the top layer, but horizontal hydraulic conductivity and vertical hydraulic conductivity of 0.1 m/day and 0.01 m/day, respectively.

The model is simulated with three stress periods. The first is steady state and does not include pumping. The second stress period is transient, includes pumping, and has a duration of 120 days with 20 time steps and a time step multiplier of 1.2. The third stress period is the same as the second except that pumping is turned off.

Simulated results for the parent model are shown in figure 11; the "hole" in the model is the volume occupied by the child. At early times, the drawdown from the pumping is limited to the vicinity of the well, as shown in figure 11*A*. However, this example is constructed such that the continued pumping causes the entire top layer of the child model to go dry. Furthermore, the drying propagates across the parent/child interface and causes several cells in the parent model to go dry, as shown in figure 11*B*. When the pumping ceases, the parent cells begin to rewet as flow from the eastern constant-head boundary restores the hydraulic head, as shown in figure 11*C* through *E*. Eventually, the heads are completely restored throughout both the parent and child model, as shown in figure 11*F*.

The situation simulated in this example is designed to highlight an advantage of the ghost-node method over the shared-node method in that inactive cells (in this case, caused by drying) can be used on the parent/child grid interface. Additionally, drying and wetting of cells can propagate across the grid interface. In general, rewetting or drying of cells can cause instabilities in the head solution. This can exacerbate convergence problems for the iterative coupling used in LGR, especially if rewetting or drying occurs in cells along the interface. In general, it is recommended that this situation be avoided if possible. If it cannot be avoided, then the ghost-node method can be used to simulate this situation, but the user should be aware that additional damping or adjustment of solver and/or rewetting values may be required to achieve an accurate solution.

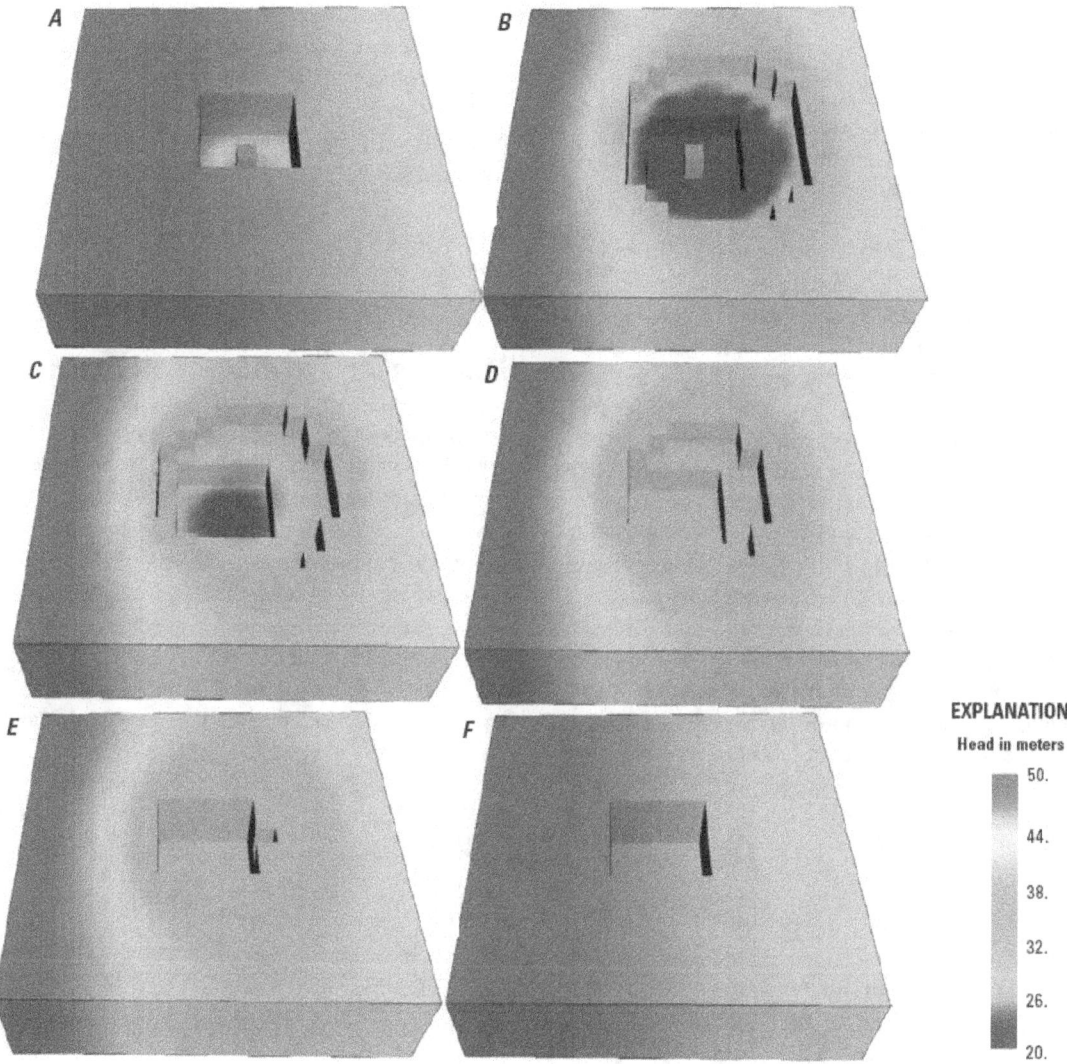

Figure 11. Simulated hydraulic heads of the parent model showing drying and rewetting of cells. The cubic-shaped hole in the parent model is the volume that is occupied by the child model and the pink column in the center is the pumping well. The larger hole is caused by cells going dry and being removed from the simulated system. Simulated results are shown after pumping for *A*, 1.2 days and *B*, 120 days; and after pumping has stopped for *C*, 0.6 days; *D*, 1.4 days; *E*, 2.3 days; and *F*, 120 days.

Example 3: Three-Dimensional Steady State Test Case with Homogeneity, Stream-Aquifer Interactions

This test case demonstrates the performance of LGR2 in a three-dimensional, unconfined aquifer. This test case also is used to illustrate the use of the Boundary Head and Flow (BFH2) Package to perform standalone simulations of either the parent or child model. The BFH2 package and inputs are described in Appendix 2.

The hypothetical groundwater model used in this analysis is shown in figure 12 and is the same three-dimensional example that was investigated in the documentation of the shared-node method (Mehl and Hill, 2005) and by Mehl and Hill (2010) for investigating grid-size dependency of streambed conductances. The meandering stream has a total length of 3,409 m and has a linear drop in stage along the length of the river from the inlet at 50.0 m to the outlet at 45.0 m. This results in a gradient along the river of 0.00147. The width, thickness of the streambed, and the streambed hydraulic conductivity are 1.0 m, 0.5 m, and 1.0 m/day, respectively. These values are constant throughout the entire stream length. The land-surface elevation of the model domain follows a linear profile from 50 m at the left boundary and drops to 45 m at the right boundary. The bottom elevation also follows this linear profile such that the model has a uniform thickness of 50 m throughout the domain. The specified-head boundaries at both ends provide a background gradient equal to the slope of the top and bottom of the model (0.00347). The aquifer is homogeneous and isotropic with a hydraulic conductivity of 1.0 m/day. The system is unconfined, which causes nonlinearity in the flow because the saturated thickness depends on the value of head, which is not known beforehand.

The parent grid is 15×15×3 and the child model is a 3:1 ratio of refinement, which results in a 15×18×6 child grid, as shown in figure 12. The refinement also extends in the vertical direction from the top of the model down to the bottom of the second layer of the three-layer parent model. Thus, in the refined region, a single parent cell is replaced by 27 child cells.

Convergence and Analysis of Flux Errors

The goal of local grid refinement is to approach the accuracy of a globally refined grid. Thus, for this analysis, each comparison is made to results obtained from a globally refined model—a model with grid spacing equivalent to the child grid over the entire domain. This example is used to analyze how the accuracy of the refined grid solution changes with each LGR iteration. Because stream/aquifer interactions are of interest, comparisons of river leakage within the child-model area at each iteration are used as the basis for comparison and are presented in figure 13. The percent difference in the ghost-node fluxes, which is a measure of the quality of the

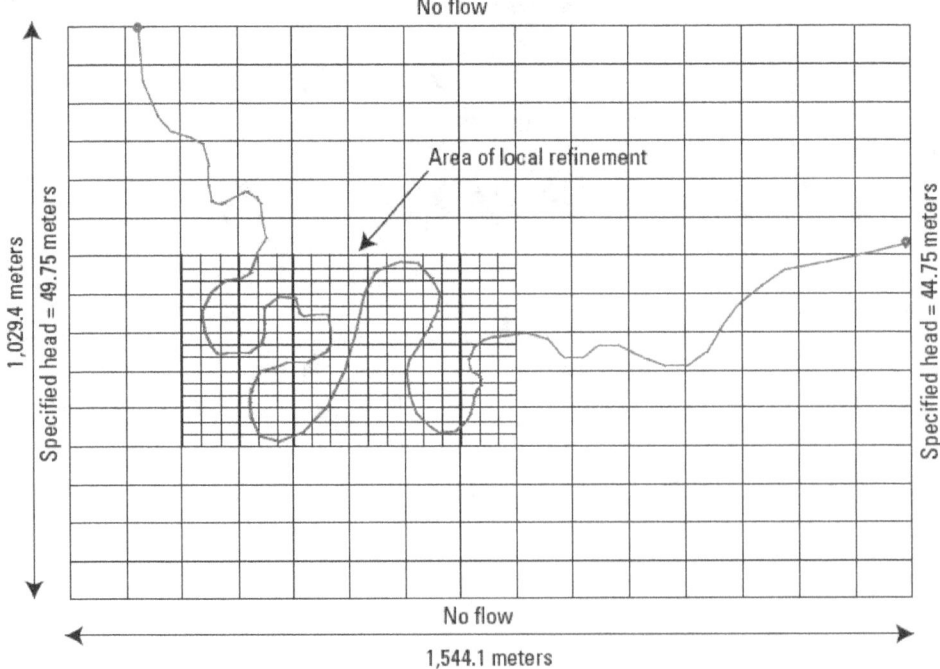

Figure 12. Plan view of a three-dimensional aquifer system used to test the local grid refinement method. A 15×15 horizontal grid discretization is shown for the parent grid and the locally refined grid (15×18) spacing is equivalent to a 45×45 discretization over the whole domain.

LGR2 solution, is also included in figure 13. Average percent differences in cell-by-cell river leakages between the globally refined solution and the one-way coupled and iteratively coupled solutions of the child model are listed in table 4.

The results in figure 13 are similar to results obtained using the shared node method (Mehl and Hill, 2005), in that a converged solution must satisfy head and fluxes on both the parent and the child grid and the errors oscillate in early iterations before they diminish. This oscillation results from the iterative coupling between heads and fluxes—it tends to update the heads and fluxes in the correct direction, but overshoots. The relaxation applied between successive iterations keeps the errors from growing, and the oscillations eventually diminish. In this case, the solution stabilizes after 10 iterations, and further iterations do not substantially improve the solution. Interestingly, for this test case, the rate of river leakage on the fourth LGR iteration has the closest match to the globally refined grid solution, as shown in figure 13. However, subsequent LGR iterations produce river leakages that are lower and farther from the globally refined solution.

In the iterative procedure used in LGR2, the errors in head and flux influence each other until a solution is achieved where the two grids are in equilibrium. At this point, the parent and child grids are in equilibrium at the interfacing boundary—interpolated ghost node heads based on the parent grid heads result in a child grid simulation that produces fluxes across the interfacing boundary that are consistent with the parent head solution. During the oscillatory approach to the final solution, it is possible that the heads or fluxes simulated on the child grid will have a better accuracy (in terms of matching the globally refined grid solution) than the final, converged solution, but this situation will not be known in advance.

Despite the decrease in average cell-by-cell river leakage error when using an iterative coupling versus a one-way coupling (table 4), the errors are still large enough to be of concern. This is to some degree an artifact of the grid-size dependence of river leakage, as shown by Mehl and Hill (2010). If LGR is used to simulate systems with substantial stream/aquifer interactions, recalibration of streambed properties might be required to obtain accurate results.

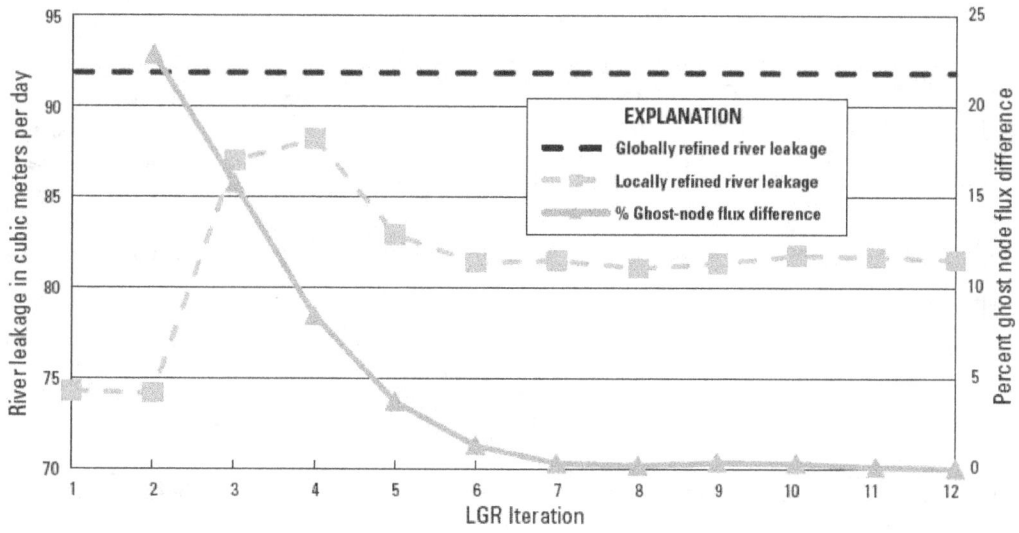

Figure 13. Total river leakage in the child model as simulated using the globally refined and locally refined models in relation to number of iterations of the local grid refinement procedure are shown with dashed lines. Ghost node flux differences in relation to the number of iterations are shown with a solid line.

Table 4. Average percent difference in river leakages in the child model using a one-way coupled and iteratively coupled solutions compared to the globally refined solution.

[%, percent]

Methods compared	Average river leakage error (%)
One-way	-15.9
Iterative	4.86

Example 4: Multiple Refined Areas

A two-dimensional example is used to illustrate the multiple-refined-areas capability. Hydraulic head contours from this example are shown in figure 14. This example is a modification of Example 1. The parent model grid cells are 9.25 m by 9.0 m, which produces a grid that has 50 rows and 108 columns. The grid cells for both child models are 1.028 m by 1.0 m, representing a 9:1 refinement ratio. The modifications include the addition of a second pumping well of identical pumping rate (5.5×10^{-3} m³/s), a homogeneous hydraulic conductivity field of 5.0×10^{-4} meters per second (m/s), and constant-head boundaries of 10.0 m on the left and right sides. This test system was designed to have a symmetric flow field such that both child models have identical solutions while simulating different parts of the domain. This forms a test of the LGR program and, indeed, the solutions in the two grids were the same.

Table 5 shows a comparison of the simulated hydraulic head at the wells and the computer processing times required for: (1) the coarse grid only, (2) the coarse grid coupled to the locally refined grids by using LGR, and (3) a globally refined grid. The globally refined grid uses the same grid spacing over the entire domain as the locally refined grids. The coarse grid coupled to the local grid by using LGR simulates a hydraulic head at the well that is in good agreement with the globally refined grid and requires somewhat less computer processing time. Approximately half the computer processing time (2.32 seconds) is required if only one area of refinement is simulated. This example demonstrates that the locally refined grids can improve the simulated drawdowns near the pumping wells (compared to the coarse grid), which is expected, and the approximately linear scaling of computer processing time when additional areas are simulated.Though the execution time is only reduced slightly in this problem with the two local grids, refining multiple areas in larger models is likely to be more computationally advantageous. Local refinement allows the rest of the model to remain unchanged, and it also allows each of the models (here, the parent model or either of the child models) to be simulated independently, which can be advantageous when simulating computationally demanding transport processes.

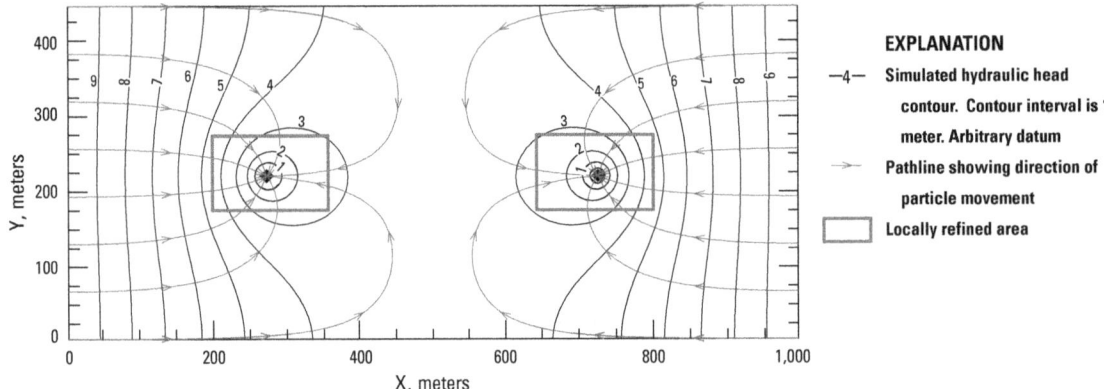

Figure 14. Example of grid refinement around two areas of pumping within a coarse-grid model. The system was designed to have a symmetric flow field.

Table 5. Comparison of simulated hydraulic head at the wells and computer processing time for several grids. Only one well is reported because the flow field is symmetric, as shown in figure 14.

[Computation times using a Linux workstation, Pentium IV—3.0 gigahertz]

Grid	Head at wells (meters)	Computer processing time (secods)
Coarse	-3.111	0.070
Locally refined	-6.959	4.113 (2.32)[1]
Globally refined	-6.957	4.164

[1]With one area of refinement

References Cited

Anderman, E.R., Kipp, K.L., Hill, M.C., Valstar, J., and Neupauer, R.M., 2002, MODFLOW-2000, the U.S. Geological Survey modular ground-water model—Documentation of the Model-Layer Variable-Direction Horizontal Anisotropy (LVDA) capability of the Hydrogeologic-Unit Flow (HUF) Package: U.S. Geological Survey Open-File Report 2002–409, 61 p. (Also available at *http://water.usgs.gov/nrp/gwsoftware/modflow2000/modflow2000.html.*)

Davison, R.M., and Lerner, D.N., 2000, Evaluating natural attenuation of groundwater pollution from a coal-carbonisation plant—Developing a local-scale model using MODFLOW, MODTMR and MT3D: Water and Environmental Management Journal, v. 14, p. 419–426.

de Marsily, G., 1986, Quantitative hydrogeology: Orlando, Fla., Academic Press, Inc., 440 p.

Dickinson, J.E., James, S.C., Mehl, S.W., Hill, M.C., Leake, S.A., Zyvoloski, G.A., Faunt, C.C., Eddebbarh, A.A., 2007, New ghost-node method for linking different models with varied grid refinement and initial investigations of heterogeneity and nonmatched grids: Advances in Water Resources, v. 30, p.1722-1736, doi:10.1016/j.advwatres.2007.01.004.

Dickinson, J.E., Hanson, R.T., Mehl, S.W., and Hill, M.C., 2011, Particle tracking in locally refined grids using MODFLOW-LGR and MODPATH-LGR: U.S. Geological Survey Techniques and Methods 6-A38, 42 p. (Also available at *http://pubs.usgs.go-v/tm/tm6a38/*)

Doherty, J., 2010, PEST, model independent parameter estimation—User manual (5th ed.): PEST, accessed January 18, 2013, at http://www.pesthomepage.org/.

Edwards, M.G., 1999, A high-resolution method coupled with local grid refinement for three-dimensional aquifer remediation: In Situ, v. 23, no. 4, p. 333–377.

Ewing, R.E., Lazarov, R.D., and Vassilevski, P.S., 1991, Local refinement techniques for elliptic problems on cell-centered grids—I. Error analysis: Mathematics of Computation, v. 56, no. 194, p. 437–461.

Funaro, D., Quarteroni, A., and Zanolli, P., 1988, An iterative procedure with interface relaxation for domain decomposition methods: Siam Journal on Numerical Analysis, v. 25, no. 2, p. 1213–1236.

Garcia, J.E., 1995, An experimental investigation of upscaling of water flow and solute transport in saturated porous media: Boulder, Colo., University of Colorado at Boulder, MS Thesis, 135 p.

Haefner, F., and Boy, S., 2003, Fast transport simulation with an adaptive grid refinement: Ground Water, v. 41, no. 2, p. 273–279.

Harbaugh, A.W., 1995, Direct solution package based on alternating diagonal ordering for the U.S. Geological Survey Modular Finite-Difference Ground-Water Flow Model: U.S. Geological Survey Open-File Report 95–288, 46 p. (Also available at *http://water.usgs.gov/nrp/gwsoftware/modflow2000/modflow2000.html.*)

Harbaugh, A.W., 2005, MODFLOW-2005, the U.S. Geological Survey modular ground-water model—The Ground-Water Flow Process: U.S. Geological Survey Techniques and Methods 6-A16, chap. 9, 62 p.

Harbaugh, A.W., Banta, E.R., Hill, M.C., and McDonald, M.G., 2000, MODFLOW-2000, the U.S. Geological Survey modular ground-water model—User guide to modularization concepts and the ground-water flow process: U.S. Geological Survey Open-File Report 00–92, 121 p.

Heywood, C.E., 2013, Simulations of groundwater flow, transport, and age in Albuquerque, New Mexico, for a study of transport of anthropogenic and natural contaminants (TANC) to public-supply wells, U.S. Geological Survey Scientific Investigations Report 2012-5242.

Hill, M.C., 1990, Preconditioned Conjugate-Gradient 2 (PCG2), A computer program for solving ground-water flow equations: U.S. Geological Survey Water-Resources Investigations Report 90–4048, 43 p. (Also available at *http://water.usgs.gov/nrp/gwsoftware/modflow2000/modflow2000.html.*)

Hunt, R.J., Steuer, J.J., Mansor, M.T.C., and Bullen, T.D., 2001, Delineating a recharge area for a spring using numerical modeling, Monte Carlo techniques, and geochemical investigation: Ground Water, v. 39, no. 5, p. 702–712.

Leake, S.A., Lawson, P.W., Lilly, M.R., and Claar, D.V., 1998, Assignment of boundary conditions in embedded ground water flow models: Ground Water, v. 36, no. 4, p. 621–625.

Leake, S.A., and Claar, D.V., 1999, Procedure and computer programs for telescopic mesh refinement using MODFLOW: U.S. Geological Survey Open-File Report 99–238, 53 p., accessed January 18, 2013 at *http://az.water.usgs.gov/MODTMR/tmr.html.*

Mapa, R., Illangasekare, T.H., and Garcia, J.E., 1994, Upscaling of water flow and solute transport in saturated porous media—theory, computation and experiments: Progress report submitted to U.S. Army Waterways Experiment Station, 184 p.

Matott, L.S., 2005, OSTRICH, An optimization software tool, documentation and user's guide, Version 1.6: Buffalo, N.Y., State University of New York at Buffalo, 114 p., accessed January 31, 2006 at *http://www.groundwater.buffalo.edu/software/Ostrich/OstrichMain.html.*

McDonald, M.G., and Harbaugh, A.W., 1988, A modular three-dimensional finite-difference ground-water flow model: U.S. Geological Survey Techniques of Water-Resources Investigations, book 6, chap. A1, 548 p. (Also available at *http://water.usgs.gov/pubs/twri/twri6a1/.*)

Mehl, S., and Hill, M.C., 2001, MODFLOW-2000, The U.S. Geological Survey modular ground-water model—User guide to the Link-AMG (LMG) package for solving matrix equations using an algebraic multigrid solver: U.S. Geological Survey Open-File Report 01–177, 33 p. (Also available at *http://water.usgs.gov/nrp/gwsoftware/modflow2000/modflow2000.html.*)

Mehl, S.W., and Hill, M.C., 2005, MODFLOW-2005, the U.S. Geological Survey modular ground-water model—Documentation of shared node local grid refinement (LGR) and the boundary flow and head (BFH) package: U.S. Geological Survey Techniques and Methods 6–A12, 68 p. (Also available at *http://water.usgs.gov/nrp/gwsoftware/modflow2005_lgr/mflgr.html.*)

Mehl, S.W., and Hill, M.C., 2007, MODFLOW-2005, the U.S. Geological Survey modular ground-water model—Documentation of multiple-refined area capability of local grid refinement (LGR) and the boundary flow and head (BFH) package: U.S. Geological Survey Techniques and Methods 6–A21, 13 p. (Available at *http://water.usgs.gov/nrp/gwsoftware/modflow2005_lgr/mflgr.html.*)

Mehl, S., and Hill, M.C. 2010, Grid-size dependence of Cauchy boundary conditions used to simulate stream-aquifer interactions: Advances in Water Resources, v. 33, no. 4, p. 430–442.

Nacul, E.C., 1991, Use of domain decomposition and local grid refinement in reservoir simulation: Stanford, Calif., Stanford University, Ph.D. thesis, 370 p.

Naff, R.L., Banta, E.R., 2008, The U.S. Geological Survey modular ground-water Model—PCGN: A preconditioned conjugate gradient solver with improved nonlinear control: U.S. Geological Survey Open-File Report 2008-1331, 35 p. (Available at *http://pubs.usgs.gov/of/2008/1331/*)

Poeter, E., Hill, M.C., Banta, E.R., Mehl, S., and Christensen, S., 2005, UCODE_2005 and six other computer codes for universal sensitivity analysis, calibration, and uncertainty evaluation: U.S. Geological Survey Techniques and Methods 6-A11, 283 p. (Also available at *http://www.mines.edu/igwmc/freeware/ucode/.*)

Pollock, D.W., 1994, User's guide for MODPATH/MODPATH-PLOT, Version 3—A particle tracking post-processing package for MODFLOW, the U.S. Geological Survey finite-difference ground-water flow model: U.S. Geological Survey Open-File Report 94–464, 249 p. (Also available at *http://water.usgs.gov/nrp/gwsoftware/modpath41/modpath41.html.*)

Quandalle, P., and Besset, P., 1985, Reduction of grid effects due to local sub-gridding in simulations using a composite grid, *in* SPE Symposium on Reservoir Simulation, 8, Dallas, Tex., Feb. 10–13, 1985, Proceedings: Richardson, Tex., Society of Petroleum Engineers, Paper SPE 13527, p. 295–305. Schaars, F., and Kamps, P., 2001, MODGRID—Simultaneous solving of different ground-water flow models at various scales, *in* MODFLOW 2001 and Other Modeling Odysseys Conference, Golden, Colo., Sep. 11–14, 2001, Proceedings: Golden, Colo., Integrated Groundwater Modeling Center, Colorado School of Mines, vol. I, p. 38–44.

Székely, F., 1998, Windowed spatial zooming in finite-difference ground water flow models: Ground Water, v. 36, no.5, p. 718–721.

Vilhelmsen, T.N., Christensen, C., and Mehl, S., 2011, Evaluation of two versions of MODFLOW-LGR to simulate regional-scale groundwater flow in a synthetic buried valley aquifer system: Ground Water, v. 50, no.1, p. 118–132, doi: 10.1111/j.1745-6584.2011.00826.x.

Ward, D.S., Buss, D.R., Mercer, J.W., and Hughes, S.S., 1987, Evaluation of a groundwater corrective action at the Chem-Dyne Hazardous Waste Site using a telescopic mesh refinement modeling approach: Water Resources Research, v. 23, no. 4, p. 603–617.

Wasserman, M.L., 1987, Local grid refinement for three-dimensional simulators, *in* SPE Symposium on Reservoir Simulation, 9, San Antonio, Tex., Feb. 1–4, 1987, Proceedings: Richardson, Tex., Society of Petroleum Engineers, Paper SPE 16013, p. 231–241.

Wilson, J.D., and Naff, R.L., 2004, The U.S. Geological Survey modular ground-water model—GMG linear equation solver package documentation: U.S. Geological Survey Open-File Report 2004–1261, 47 p. (Also available at *http://water.usgs.gov/pubs/of/2004/1261/.*)

Zheng, C., and Wang, P., 1999, MT3DMS—A modular three-dimensional multi-species transport model for simulation of advection, dispersion and chemical reactions of contaminants in groundwater systems—Documentation and user's guide—Contract report SERDP-99-1: Vicksburg, Miss., U.S. Army Engineer Research and Development Center, 202 p. (Available at *http://hydro.geo.ua.edu/mt3d.*)

Appendixes

Appendix 1.   LGR2 Input Instructions and Selected Input and Output Files from Examples 1 and 3

LGR2 Input Instructions

When executed, MODFLOW-2005 prompts for the name of a file. If a Name File (Harbaugh and others, 2000, p. 7, 43) is entered, LGR is not used. To use LGR, version 2.0, (LGR2), the name of the LGR2 Control File is entered. The contents of this file are described here.

The LGR2 Control File is distinguished from a Name file by the presence of a keyword "LGR" as the first non-commented input. LGR2 reads its input data from this file. Input for LGR2 is defined using 15 items. Each item is read free format.

FOR EACH SIMULATION

1. LGR

2. NGRIDS

FOR THE PARENT GRID (the parent grid needs to be listed before the child grid)

3. NAMEFILE

4. GRIDSTATUS

5. IUPBHSV IUPBFSV

FOR EACH CHILD GRID [repeat items 6 through 15 for each grid with the exception of the parent grid. That is, repeat these items a total of (NGRIDS - 1) times]

6. NAMEFILE

7. GRIDSTATUS

8. ISHFLG IBFLG IUCBHSV IUCBFSV

9. MXLGRITER IOUTLGR

10. RELAXH RELAXF

11. HCLOSELGR FCLOSELGR

12. NPLBEG NPRBEG NPCBEG

13. NPLEND NPREND NPCEND

14. NCPP

15. NCPPL [Repeat NCPPL a total of (NPLEND +1 - NPLBEG) times]

Explanation of Variables Read by LGR2

NGRIDS—is the total number of grids used in this simulation, including the parent grid and all of the child grids.
NAMEFILE— is the name of the Name file for either the parent or child grid. The name can include the file path and is limited to 200 characters.
GRIDSTATUS—is a character variable indicating whether the file listed in NAMEFILE corresponds to a parent or child grid.
    If GRIDSTATUS = PARENTONLY, then it is a parent grid Name file.
    If GRIDSTATUS = CHILDONLY, then it is a child grid Name file.
IUPBHSV—a number greater than zero that corresponds to the unit number where the boundary heads are saved for later use by the Boundary Flow and Head (BFH2) Package for independent simulations. A file with this unit number needs to be opened in the Name file of the parent model. A value of zero indicates that the file is not written. For the parent model, these are the complementary boundary conditions (see Appendix 2).

ISHFLG—is a flag indicating whether heads from the parent grid-simulation should be used as the starting head for the child grid simulation. These heads apply to the interior of the child, not the boundary.

 If ISHFLG = 1, then use results of the parent-grid simulation as the starting head for the child grid. In the cells of the child grid that overlap the parent grid, the heads of the corresponding parent cell are used. No interpolation is applied. For steady-state simulations, this can provide a good initial guess which can reduce computational time. For transient simulations, this overwrites the initial condition of the child model defined in STRT of the Basic Package input file (Harbaugh, 2005) and therefore is not recommended.

 If ISHFLG = 0, then use the heads defined in STRT of the Basic Package for the child grid.

IBFLG—is a positive integer used to define the interface of the child grid with the parent. Use this value around the perimeter of the child model IBOUND array. Do not use IBFLG or -IBFLG anywhere else in the parent or child IBOUND arrays. Use a unique value for each child grid.

IUCBHSV—a number greater than zero that corresponds to the unit number where the boundary heads are saved for later use by the BFH2 Package for independent simulations. A file with this unit number needs to be opened in the Name file of the child model. A value of zero indicates that the file is not written. For the child model, these are the coupling boundary conditions (see Appendix 2).

IUCBFSV—a number greater than zero that corresponds to the unit number where the boundary fluxes are saved for later use by the BFH2 Package for independent simulations. A file with this unit number needs to be opened in the Name file of child model. A value of zero indicates that the file is not written. For the child model, these are the complementary boundary conditions (see Appendix 2).

MXLGRITER—is the maximum number of LGR iterations; 20 iterations are sufficient for most problems. See Closure Criteria for LGR Iterations section. Set MXLGRITER to 1 for a one-way coupling.

IOUTLGR—is a flag that controls printing from LGR iterations of the maximum head and flux change. For the maximum head change, the head value and corresponding layer, row, and column of the child grid is listed. For the maximum flux change, the flux value and corresponding layer, row, and column of the parent grid is listed. If IOUTLGR < 0, output is written to the screen. If IOUTLGR >0, output is written to the child listing file. If IOUTLGR = 0, no results are written.

RELAXH—is the relaxation factor for heads.

RELAXF—is the relaxation factor for fluxes and flag for using method 1 or 2 formulation.

 IF RELAXF > 0 and < 1.0, method 1 is used to formulate the coupling equations (see section "Two Methods of Formulating and Solving the Equations"). In this case, values of RELAXH and RELAXF less than 1 and greater than 0 are needed for convergence of the LGR iterations. Typically, values around 0.5 produce convergent solutions. Values less than 0.5 may be needed when the LGR iterations have difficulty converging. In cases when the LGR iterations exhibit no convergence difficulties, values greater than 0.5 may reduce the number of iterations needed or convergence. Convergence difficulties can be diagnosed by printing the maximum head and flux changes (IOUTLGR $\neq$ 0) to determine if the head and flux changes are decreasing (converging) or increasing (diverging) as the LGR iterations proceed.

 IF RELAXF $\leq$ 0, method 2 is used to formulate the coupling equations (see section "Two Methods of Formulating and Solving the Equations"). In this formulation only

 RELAXH is used. Typically, a value of RELAXH = 1.0 produces convergent solutions. Values of RELAXH greater than 1.0 and less than 1.5 may reduce the number of LGR iterations.

HCLOSELGR—is the head closure criterion for the LGR iterations. The closure criterion is based on heads of the child interface nodes. This closure criterion is satisfied when the maximum absolute head change between successive LGR iterations is less than HCLOSELGR (see equation 15b).

FCLOSELGR—is the flux closure criterion for the LGR iterations. The closure criterion is based on fluxes into the parent interface nodes. This closure criterion is satisfied when the maximum absolute relative flux change between successive LGR iterations is less than FCLOSELGR (see equation 15a).

NPLBEG—is the number of the topmost layer of the parent grid where the child model begins. Currently (2013), refinement must begin at the top of the model so NPLBEG = 1.

NPRBEG—is the row number of the parent grid where the refinement begins.

NPCBEG—is the column number of the parent grid where the refinement begins.

NPLEND—is the number of the lowest layer of the parent grid where the refinement ends. NPLEND $\geq$ NPLBEG

NPREND—is the row number of the parent grid where the refinement ends. NPREND > NPRBEG.

NPCEND—is the column number of the parent grid where the refinement ends. NPCEND > NPCBEG.

NCPP—is the number of child cells that span the width of a single parent cell along rows and columns. This must be an even or odd integer number > 1 and is applied to rows and columns.

NCPPL—is the number of child cells that span the depth of a single parent layer. This must be an even or odd integer number $\geq$ 1. Read one value for each refined parent layer. The number of values needs to equals NPLEND +1 minus NPLBEG. Values can be 1, which results in no vertical refinement for the layer; for example, in figure 4c the values 1 3 3 would be needed.

Example LGR2 Input Files

The sample data inputs listed below are for the two-dimensional example 1 and three-dimensional example 3 presented in the text of this report. The systems are shown in figure 8 and figure 12.

Example 1 uses a 9:1 refinement ratio. The refinement begins in layer 1, row 20, column 23 and ends in layer 1, row 31, column 38 of the parent grid. The annotated LGR input file is:

```
LGR                   #LGR Keyword
2                     #NGRIDS
ex2_parent nam        #NAME FILE
PARENTONLY            #GRIDSTATUS
70 71                 #Unit numbers for complementary and coupling boundaries
ex2_child nam         #NAME FILE
CHILDONLY             #GRIDSTATUS
1 59 80 81            #ISHFLG, IBFLG, Unit numbers for complementary and coupling boundaries
15 0                  #MXLGRITER, IOUTLGR
0.50 0.50             #RELAXH, RELAXF
1.0E-5 1.0E-5         #HCLOSELGR, FCLOSELGR
1 20 23               #Beginning layer, row, and column
1 31 38               #Ending layer, row, and column
9                     #Horizontal refinement ratio
1                     #Vertical refinement ratio by Parent layer
```

Example 3 is for a three-dimensional problem and uses a 3:1 ratio of refinement in all directions, in the top two layers of the three-layer parent model. The refinement begins in layer 1, row 7, column 3 and ends in layer 2, row 11, column 8 of the parent grid (see figure 12).

```
LGR                   #LGR Keyword
2                     #NGRIDS
ex3_parent nam        #NAME FILE
PARENTONLY            #GRIDSTATUS
0 0                   #Unit numbers for complementary and coupling boundaries.
ex3_child nam         #NAME FILE
CHILDONLY             #GRIDSTATUS
1 59 80 81            #ISHFLG, IBFLG, Unit numbers for complementary and coupling boundaries
15 0                  #MXLGRITER, IOUTLGR
0.40 0.40             #RELAXH, RELAXF
5.0E-3 5.0E-2         #HCLOSELGR, FCLOSELGR
1 7 3                 #Beginning layer, row, and column
2 11 8                #Ending layer, row, and column
3                     # Horizontal refinement ratio
3 3                   # Vertical refinement ratio by Parent layer
```

Sample LGR2 Output

The output from MODFLOW-2005 when using LGR differs from standard MODFLOW-2005 output in two ways. First, the LGR control input file is echoed. Second, the volumetric budget includes ghost-node fluxes through the ghost-node heads at the interface between grids. When LGR is active, the child grid output includes an additional volumetric budget that accumulates the ghost-node fluxes in and out from the parent and child grids and calculates the percent difference between them. This can be used to assess the quality of the LGR2 solution (see the "Closure Criteria for LGR Iterations" section). Using the three-dimensional example 3, the output appears as follows:

Parent model output:

•••

```
LGR2 — LOCAL GRID REFINEMENT, VERSION 2.0, 06/25/2013
   INPUT READ FOR MODEL 1 DEFINED BY NAME FILE ex3_parent.nam
   LOCAL GRID REFINEMENT IS ACTIVE FOR PARENT ONLY
```

•••

VOLUMETRIC BUDGET FOR ENTIRE MODEL AT END OF TIME STEP 1 IN STRESS PERIOD 1

CUMULATIVE VOLUMES	L**3	RATES FOR THIS TIME STEP	L**3/T
IN:		IN:	
—		—	
STORAGE =	0.0000	STORAGE =	0.0000
CONSTANT HEAD =	182.5212	CONSTANT HEAD =	182.5212
RIVER LEAKAGE =	152.7981	RIVER LEAKAGE =	152.7981
GHOST-NODE FLUX =	187.0215	GHOST-NODE FLUX =	187.0215
TOTAL IN =	522.3408	TOTAL IN =	522.3408
OUT:		OUT:	
—		—	
STORAGE =	0.0000	STORAGE =	0.0000
CONSTANT HEAD =	155.3137	CONSTANT HEAD =	155.3137
RIVER LEAKAGE =	36.6952	RIVER LEAKAGE =	36.6952
GHOST-NODE FLUX =	330.3314	GHOST-NODE FLUX =	330.3314
TOTAL OUT =	522.3403	TOTAL OUT =	522.3403
IN - OUT =	4.8828E-04	IN - OUT =	4.8828E-04
PERCENT DISCREPANCY =	0.00	PERCENT DISCREPANCY =	0.00

Child model output:

...

LGR2 — LOCAL GRID REFINEMENT, VERSION 2.0, 06/25/2013
      INPUT READ FOR MODEL 2 DEFINED BY NAME FILE ex3_child nam
LOCAL GRID REFINEMENT IS ACTIVE FOR CHILD ONLY
      STARTING HEADS FROM PARENT WILL BE USED: ISHFLG = 1
      VALUE IN IBOUND INDICATING BOUNDARY INTERFACE = 59
      BOUNDARY HEADS WILL BE SAVED ON UNIT 80
      BOUNDARY FLUXES WILL BE SAVED ON UNIT 81
      MAX NUMBER OF LGR ITERATIONS = 15
      LGR ITERATIONS RESULTS NOT WRITTEN: IOUTLGR= 0

      WEIGHTING FACTORS FOR RELAXATION
      RELAXH(HEAD) RELAXF(FLUX)
      ————————————————

      0.400E+00   0.400E+00

      CLOSURE CRITERIA FOR LGR ITERATIONS
      HCLOSELGR   FCLOSELGR
      ————————————————

      5.000E-03   5.000E-02

      STARTING LAYER, ROW, COLUMN= 1, 7, 3
      ENDING LAYER, ROW, COLUMN= 2, 11, 8
      NCPP: NUMBER OF CHILD CELLS PER WIDTH OF PARENT CELL= 3
      NCPPL: NUMBER OF CHILD LAYERS IN LAYER 1 OF PARENT = 3
      NCPPL: NUMBER OF CHILD LAYERS IN LAYER 2 OF PARENT = 3

...

VOLUMETRIC BUDGET FOR ENTIRE MODEL AT END OF TIME STEP 1 IN STRESS PERIOD 1

CUMULATIVE VOLUMES	L**3	RATES FOR THIS TIME STEP	L**3/T
IN:		IN:	
—		—	

STORAGE =	0.0000	STORAGE =	0.0000
CONSTANT HEAD =	0.0000	CONSTANT HEAD =	0.0000
RIVER LEAKAGE =	81.7762	RIVER LEAKAGE =	81.7762
GHOST-NODE HEAD =	330.7134	GHOST-NODE HEAD =	330.7134
TOTAL IN =	412.4897	TOTAL IN =	412.4897

OUT: OUT:

STORAGE =	0.0000	STORAGE =	0.0000
CONSTANT HEAD =	0.0000	CONSTANT HEAD =	0.0000
RIVER LEAKAGE =	224.5170	RIVER LEAKAGE =	224.5170
GHOST-NODE HEAD =	187.9724	GHOST-NODE HEAD =	187.9724
TOTAL OUT =	412.4894	TOTAL OUT =	412.4894
IN - OUT =	2.4414E-04	IN - OUT =	2.4414E-04
PERCENT DISCREPANCY =	0.00	PERCENT DISCREPANCY =	0.00

...

FLUX ACROSS PARENT-CHILD INTERFACE AT TIME STEP 1 IN STRESS PERIOD 1

G-N FLUX	PARENT	CHILD	DIFFERENCE	% DIFFERENCE
RATE IN:	3.3033E+02	3.3071E+02	-3.8202E-01	-0.1156
RATE OUT:	1.8702E+02	1.8797E+02	-9.5094E-01	-0.5072

Reference Cited

Harbaugh, A.W., Banta, E.R., Hill, M.C., and McDonald, M.G., 2000, MODFLOW-2000, the U.S. Geological Survey modular ground-water model—User guide to modularization concepts and the ground-water flow process: U.S. Geological Survey Open-File Report 00–92, 121 p.

Appendix 2.   Independent Simulations Using the Boundary Flow and Head (BFH2) Package

The Boundary Flow and Head (BFH2) Package reads input data from the file indicated in the Name file as described by Harbaugh and others (2000, p. 7, 43) using the File Type BFH. Input for the BFH2 Package is created by LGR2 and requires that the coupling boundary conditions calculated by LGR2 be saved using variable IUCBHSV and (or) IUPBFSV of the LGR2 input file. To generate the file required for a subsequent independent child model simulation, IUCBHSV needs to be nonzero; to generate the file required for a subsequent independent parent model simulation, IUPBFSV needs to be nonzero.

The BFH2 Package and LGR2 cannot be used simultaneously. Thus, when using LGR2, the Name file specified in the LGR2 control file cannot use file type BFH.

The procedure needed to run independent child or parent models with LGR boundary conditions is as follows:

1.   Use LGR2 to calculate and save the coupling boundary conditions.

2.   Activate the BFH2 Package in the Name file with a file name that corresponds to the file saved on IUCBHSV or IUPBFSV for child and parent simulations, respectively.

As discussed in the "Running the Parent and Child Model Independently Using the Boundary Flow and Head (BFH2) Package" section, the BFH2 Package can be used to evaluate the effects of model changes on the boundary conditions. In this case, the complementary boundary conditions also need to be saved when running LGR2. For the child model, IUCBFSV needs be nonzero; for the parent model, IUPBHSV needs to be nonzero. If the file containing the complementary boundary conditions for the child or parent models is opened in the Name file on the unit number corresponding to IUCBFSV or IUPBHSV, respectively, then the BFH2 Package will evaluate the discrepancies in the complementary boundary conditions.

Each of these files contains a header record and a list of the child and parent cells involved in the coupling, indicated by the layer, row, and column. For the child models, the corresponding adjoining parent cells and a node index is listed with each child cell. This is followed by a listing of the boundary head or flux values, corresponding to these cells, for each time step.

Example BFH2 Inputs

The options for the BFH2 Package can be controlled through inputs to LGR2 and the Name files. Using the three-dimensional example 3 in Appendix 1, a simulation using LGR2 is performed first. For an independent simulation of the child grid, the coupling boundary condition (specified head) is saved on unit 80 and the complementary boundary condition (boundary flux) is saved on unit 81, as denoted by the shading in the example below.

The Name file for the child grid for the LGR2 simulation is:

```
LIST 26 ex3_child.out
BAS6 2 ex3_child.ba6
BCF6 21 ex3_child.bc6
DIS 29 ex3_child.dis
OC 20 ex3_child.oc
DATA(BINARY) 31 ex3_child hed
DATA(BINARY) 41 ex3_child.flw
PCG 22 ex3_child_3.pcg
RIV 25 ex3_child riv
DATA 80 ex3_child_bfh hed
DATA 81 ex3_child_bfh.flw
DATA 51 ex3_child.bot
```

After successful completion of an LGR2 simulation, the child model can be simulated independently using the BFH2 Package. Only the Name file of the child grid needs to be modified. Activate BFH2 with a file name corresponding to the file where the coupling boundary conditions were saved. Although not required in the example above, the complementary boundary conditions were saved. If this file is opened in the Name file on the same unit number on which it was saved, the BFH2 Package will report any changes in the boundary fluxes of the child model. This is done in the example below. Use of # in the first column results in the line being ignored. The required changes to the name file are denoted with shading.

```
LIST 26 ex3_child_bfh.out
BAS6 2 ex3_child.ba6
BCF6 21 ex3_child.bc6
```

```
DIS 29 ex3_child.dis
OC 20 ex3_child.oc
DATA(BINARY) 31 ex3_child hed
DATA(BINARY) 41 ex3_child.flw
PCG 22 ex3_child_3.pcg
RIV 25 ex3_child riv
BFH2 80 ex3_child_bfh hed
#DATA 80 ex3_child_bfh hed
DATA 81 ex3_child_bfh.flw
DATA 51 ex3_child.bot
```

Sample BFH2 Output

The output from MODFLOW-2005 when using the BFH2 Package will show the volumetric budget contributions from the coupling boundary conditions. If the complementary boundary conditions are saved by LGR2 and activated in the Name file, then the BFH2 Package reports changes in the complementary boundary conditions. That is, for the child model, which is coupled using specified-head boundary conditions, changes in flow through the interfacing boundary are reported. For the parent model, which is coupled using specified-flux boundary conditions, changes in head along the interfacing boundary are reported. Using the three-dimensional example above, the additions to the MODFLOW-2005 output for an independent child grid simulation appear as:

•••

```
BFH2 — BOUNDARY FLOW AND HEAD PACKAGE, VERSION 2.0, 06/25/2013
          INPUT READ FROM UNIT 80
GHOST-NODE HEAD
RUNNING CHILD MODEL WITH 666 SPECIFIED HEAD BOUNDARY NODES
CHECKING AGAINST FLUX BOUNDARY CONDITIONS ON UNIT 81
```

•••

VOLUMETRIC BUDGET FOR BFH SPECIFIED HEADS AT TIME STEP 1 IN STRESS PERIOD 1

CUMULATIVE VOLUMES	L**3	RATES FOR THIS TIME STEP	L**3/T
TOTAL IN =	330.7133	TOTAL IN =	330.7133
TOTAL OUT =	187.9724	TOTAL OUT =	187.9724

BFH2: BOUNDARY FLUX COMPARISON

```
NEW TOTAL BOUNDARY FLUX =        142.740921
OLD TOTAL BOUNDARY FLUX =        142.740936
AVERAGE ABSOLUTE FLUX DIFFERENCE =      0.770578534E-08
MAXIMUM ABSOLUTE FLUX DIFFERENCE OF    -0.238418579E-06
OCCURS AT PARENT LAYER 3 ROW 7 COLUMN 5
NEW FLUX AT THIS NODE =   2.35115957
OLD FLUX AT THIS NODE =   2.35115981
```

There are some small discrepancies in the boundary fluxes even though no modifications were made to the child model. This is because a new head solution is required using the BFH2 heads as the boundary conditions for the child model, and this requires a call to the solver. Therefore, the differences are on the same order magnitude as the closure criteria of the solver.

If the child model is modified, the BFH2 Package can be used to assess the effects on the coupling boundary conditions. For example, consider changing the child model to include pumping at a rate of 9.0 cubic meters per day (m³/day) from layer 2, row 6, column 10 and simulated with the BFH2 Package. The results are shown below:

...

VOLUMETRIC BUDGET FOR BFH2 SPECIFIED HEADS AT TIME STEP 1 IN STRESS PERIOD 1

CUMULATIVE VOLUMES	L**3	RATES FOR THIS TIME STEP	L**3/T
TOTAL IN =	332.2234	TOTAL IN =	332.2234
TOTAL OUT =	186.0823	TOTAL OUT =	186.0823

BFH: BOUNDARY FLUX COMPARISON

```
NEW TOTAL BOUNDARY FLUX =      146.141113
OLD TOTAL BOUNDARY FLUX =      142.740936
AVERAGE ABSOLUTE FLUX DIFFERENCE =      0.510525936E-02
MAXIMUM ABSOLUTE FLUX DIFFERENCE OF   0.566826105
OCCURS AT PARENT LAYER 3 ROW 8 COLU-MN 6
NEW FLUX AT THIS NODE =     -0.485062867
OLD FLUX AT THIS NODE =     -1.05188894
```

Although the pumping well is located below a river node, only about 5.6 m^3/day comes from the river in the form of reduced discharge and induced recharge; the remaining 3.4 m^3/day comes from the boundaries. Of this, about 0.57 m^3/day comes from the cells adjoining the parent cell in layer 2, row 8, column 6. The child cells that correspond to this parent cell can be determined from the complementary boundary condition file where the child and corresponding parent cells are listed.

Reference Cited

Harbaugh, A.W., Banta, E.R., Hill, M.C., and McDonald, M.G., 2000, MODFLOW-2000, the U.S. Geological Survey modular ground-water model—User guide to modularization concepts and the ground-water flow process: U.S. Geological Survey Open-File Report 00–92, 121 p.

Appendix 3.   Error Propagation in LGR2

For the case where specified-head boundary conditions are used around the perimeter of the child grid (as in LGR2 in the form of ghost-node heads) and when the governing groundwater flow equation is linear, the error in the specified heads propagate into the interior of the model domain. The propagation is controlled by the governing equation for the aquifer system being modeled, minus the sink and source terms. To demonstrate this, first define the specified-head boundary condition as the true head plus some error (from grid resolution, interpolation, and so on):

$$\{h_b\} = \{h_{Tb}\} + \{e\} \tag{17}$$

where

    $\{h_b\}$ is head boundary condition,
    $\{h_{Tb}\}$ is true head at the boundary, and
    $\{e\}$ is error at the boundary

The matrix equations resulting from a finite-difference discretization can be written as:

$$[A]\{h\} = \{C \times h_b + q\} \tag{18}$$

where

    $[A]$ is the standard coefficient matrix resulting from a finite-difference discretization,
    $\{h\}$ is head in the child grid,
    C is a coefficient multiplying h_b which accounts for the conductance between the specified-head boundary condition and the aquifer, and
    $\{q\}$ is all other sink and source terms in the child grid.

Substituting equation 17 into the right-hand side of equation 18:

$$[A]\{h\} = \{C \times (h_{Tb}+e) + q\} \tag{19}$$

Reordering the right-hand side, the solution can be written as:

$$\{h\} = [A]^{-1}\{C \times h_{Tb}+q+C \times e\} \tag{20}$$

The matrix multiplication can be distributed across the terms of the right-hand side:

$$\{h\} = [A]^{-1}\{C \times h_{Tb}+q\} + [A]^{-1}\{C \times e\} \tag{21}$$

which can be written as,

$$\{h\} = \{h_T\} + \{h_e\} \tag{22}$$

where

    $\{h_T\}$ is $[A]^{-1}\{C \times h_{Tb}+q\}$, and
    $\{h_e\}$ is $[A]^{-1}\{C \times e\}$

This result stems directly from the principle of superposition where the effects of two components are added together (see Reilly and others, 1987). The first term on the right-hand side of equation 22 is the true head solution, $\{h_T\}$, which would be obtained if the true boundary conditions were used, $\{h_{Tb}\}$. The second term on the right-hand side, $\{h_e\}$, has the same coefficient matrix and represents how the additional error on the boundary is diffused through the grid. Because there are no sinks or sources for this second term (the sinks/sources are accounted for in the first term), it contains the boundary errors only. This results in two important properties: (1) the maximum error occurs at the boundary and (2) the error is propagated from the boundary through the grid by a purely diffusive process with the same coefficients of the groundwater flow system. The diffusion process causes a smoothing effect by averaging with neighboring cells. Thus, positive and negative errors on the boundary tend to cancel as they propagate into the interior. If the error is constant along the boundary, there will be no cancellation by averaging with neighboring cells, and the error is propagated directly into the interior.

The above analysis is strictly valid only when the governing groundwater flow equation is linear and the principle of super-position holds. For nonlinear situations, such as flow in unconfined aquifers, the errors are still propagated through the grid by a diffusion process; however, the coefficients in the matrix are not identical to those from the groundwater flow system.The above analysis can be used to evaluate the errors from transient simulations because the numerical solution to the transient groundwater flow equations can be viewed as a series of steady-state solutions. The error after the first time step is the same as that outlined above except that the coefficient matrix [A] and right-hand side include a storage term. The errors at subsequent time steps include diffusion of errors in the interior from the previous time step, plus the error introduced at the boundary at the current time step. Starting from equation 18, the modification for transient flow is:

$$[A]\{h^n\} = \{C \times h_b + S \times h^{n-1} + q^n\} \tag{23}$$

where,
superscripts n and $n-1$ denote the current time step and previous time step, respectively, and S is a coefficient multiplying the heads at the previous time which accounts for the changes in storage during the time step.

Substituting equation 17 for h_b and equation 22 for h^{n-1} in equation 23 and following the previous derivation:

$$\{h^n\} = [A]^{-1}\{C \times h_{Tb} + S \times h_T^{n-1} + q^n\} + [A]^{-1}\{C \times e^n + S \times h_e^{n-1}\} \tag{24}$$

The first term on the right-hand side of equation 24 is the true head solution that would be obtained if true boundary conditions were used in this time step, and the previous solution also were true. The second term represents the error due to errors at the boundary of this time step plus the errors that were propagated into the interior from the previous time step. Setting the storage term, S, to 0 in equation 24 simplifies to the steady-state solution shown in equation 21, indicating that the errors will approach those given in equation 21 as the solution approaches steady state.

Reference Cited

Reilly, T.E., Franke, O.L., and Bennett, G.D., 1987, The principle of superposition and its application in ground-water hydraulics: U.S. Geological Survey Techniques of Water-Resources Investigations, book 3, chap. B6, 28 p.

Appendix 4.   LGR2 Input File for Multiple Refined Areas

The sample data inputs listed below are for the two-dimensional example described in the "Example 4: Multiple Refined Areas" section of this report, and shown in figure 14. This example has two local areas of refinement, both with 9:1 refinement ratios. The first area of refinement begins in layer 1, row 20, column 23 and ends in layer 1, row 31, column 38 of the parent grid. The second area of refinement begins layer 1, row 20, column 71 and ends in layer 1, row 31, column 86. The annotated LGR2 input file is:

```
LGR                    #LGR Keyword
3                      #NGRIDS
parent.nam             #NAME FILE
PARENTONLY             #GRIDSTATUS
70 71                  #Unit numbers for complementary and coupling boundaries
child1.nam             #NAME FILE for the first child grid
CHILDONLY              #GRIDSTATUS
1 59 80 81             #ISHFLG, IBFLG, Unit numbers for complementary and coupling boundaries
20 0                   #MXLGRITER, IOUTLGR
0.50 0.50              #RELAXH, RELAXF
1.0E-6 1.0E-6          #HCLOSELGR, FCLOSELGR
1 20 23                #Beginning layer, row, and column
1 31 38                #Ending layer, row, and column
9                      #Horizontal refinement ratio
1                      #Vertical refinement ratio by Parent layer
child2.nam             #NAME FILE for the second child grid
CHILDONLY              #GRIDSTATUS
1 39 90 91             #ISHFLG, IBFLG, Unit numbers for complementary and coupling boundaries
20 0                   #MXLGRITER, IOUTLGR
0.50 0.50              #RELAXH, RELAXF
1.0E-6 1.0E-6          #HCLOSELGR, FCLOSELGR
1 20 71                #Beginning layer, row, and column
1 31 86                #Ending layer, row, and column
9                      #Horizontal refinement ratio
1                      #Vertical refinement ratio by Parent layer
```

LGR2 Input File Showing Available Flexibility

The sample data inputs listed below are for a three-dimensional example with two areas of refinement. This example demonstrates the available flexibility because refinement ratios and convergence criteria can be different for each child grid. The first area of refinement begins in layer 1, row 20, column 22 and ends in layer 2, row 31, column 39 of the parent grid and uses a 9:1 refinement ratio horizontally and a 5:1 ratio of refinement vertically. The second area of refinement begins layer 1, row 6, column 22 and ends in layer 2, row 17, column 39 and uses a 3:1 refinement ratio horizontally and a 4:1 and 3:1 ratio of refinement vertically for the first and second parent layers, respectively. The annotated LGR2 input file is:

```
LGR                    #LGR Keyword
3                      #NGRIDS
ex5_parent nam         #NAME FILE
PARENTONLY             #GRIDSTATUS
70 71                  #Unit numbers for complementary and coupling boundaries
ex5_child1 nam         #NAME FILE for the first child grid
CHILDONLY              #GRIDSTATUS
1 59 80 81             #ISHFLG, IBFLG, Unit numbers for complementary and coupling boundaries
25 -1                  #MXLGRITER, IOUTLGR
0.50 0.50              #RELAXH, RELAXF
1.0E-5 1.0E-5          #HCLOSELGR, FCLOSELGR
1 20 22                #Beginning layer, row, and column
2 31 39                #Ending layer, row, and column
```

```
9                         #Horizontal refinement ratio
5 5                       #Vertical refinement ratio by Parent layer
ex5_child2 nam            #NAME FILE for the second child grid
CHILDONLY                 #GRIDSTATUS
1 39 90 91                #ISHFLG, IBFLG, Unit numbers for complementary and coupling boundaries
15 0                      #MXLGRITER, IOUTLGR
0.40 0.30                 #RELAXH, RELAXF
1.0E-4 1.0E-4             #HCLOSELGR, FCLOSELGR
1 6 22                    #Beginning layer, row, and column
2 17 39                   #Ending layer, row, and column
3                         #Horizontal refinement ratio
4 3                       #Vertical refinement ratio by Parent layer
```

Appendix 5.   Brief Program Description

Table 5-1. Variables in Fortran module LGRMODULE.

Variable name	Size	Description
ISCHILD	Scalar	Flag: -1=parent grid, 1=child grid.
NGRIDS	Scalar	Number of grids used in the simulation
NPLBEG	Scalar	Layer in the parent grid where refinement begins.
NPRBEG	Scalar	Row in the parent grid where refinement begins.
NPCBEG	Scalar	Column in the parent grid where refinement begins.
NPLEND	Scalar	Layer in the parent grid where refinement ends.
NPREND	Scalar	Row in the parent grid where refinement ends.
NPCEND	Scalar	Column in the parent grid where refinement ends.
NCPP	Scalar	Refinement ratio along rows and columns.
NPL	Scalar	Number of parent layers that are refined.
IBOTFLG	Scalar	Flag: 0=bottom layer is not refined, 1=bottom layer is refined.
ISHFLG	Scalar	Flag: 0=do not use initial parent solution for interior of child, 1=use initial parent solution for interior of child.
IBFLG	Scalar	Unit number used in child IBOUND array that denotes the interface boundary.
IUPBHSV	Scalar	Unit number where parent boundary heads are saved for use with the BFH2 Package.
IUCBHSV	Scalar	Unit number where child boundary heads are saved for use with the BFH2 Package.
IUPBFSV	Scalar	Unit number where parent boundary fluxes are saved for use with the BFH2 Package.
IUCBFSV	Scalar	Unit number where child boundary fluxes are saved for use with the BFH2 Package.
MXLGRITER	Scalar	Maximum number of LGR iterations allowed.
IOUTLGR	Scalar	Flag: -1=print LGR iterations to screen, 1=print LGR iterations to listing file, 0=do not print LGR iterations.
NBNODES	Scalar	Number of child cells on the interface.
NPNODES	Scalar	Number of parent cells on the interface.
IBMAXH	Scalar	Interface cell index with maximum head change.
IBMAXF	Scalar	Interface cell index with maximum flux change.
NCMAXH	Scalar	Ghost-node connection of the interface cell with maximum head change.
NCMAXF	Scalar	Ghost-node connection of the interface cell with maximum flux change.
RELAXH	Scalar	Relaxation factor for heads.
RELAXF	Scalar	Relaxation factor for fluxes.
HCLOSELGR	Scalar	Head closure criterion for LGR iterations.
FCLOSELGR	Scalar	Flux closure criterion for LGR iterations.
HDIFFM	Scalar	Maximum absolute head change between successive LGR iterations.
FDIFFM	Scalar	Maximum absolute flux change between successive LGR iterations.
PRATIN	Scalar	Accumulation of ghost-node fluxes into parent interface cells.
CRATIN	Scalar	Accumulation of ghost-node fluxes into child interface cells.
PRATOUT	Scalar	Accumulation of ghost-node fluxes out of parent interface cells.
CRATOUT	Scalar	Accumulation of ghost-node fluxes out of child interface cells.
IBPFLG	Number of grids	A list of IBOUND flags (IBFLG) used within this grid.
IEDG	Number of child boundary cells	Row number of the interface edge.
JEDG	Number of child boundary cells	Column number of the interface edge.
NCPPL	NPL	Vertical refinement ratio for each parent layer that is refined.

Table 5-1. Variables in Fortran module LGRMODULE.—Continued

Variable name	Size	Description
NODEH	3	NODEH(n) identifies cell with maximum head change (HDIFFM): n=1—Layer number. n=2—Row number. n=3—Column number.
NODEF	3	NODEF(n) identifies cell with maximum flux change (FDIFFM): n=1—Layer number. n=2—Row number. n=3—Column number.
NCON	Number of child boundary cells	Number of ghost-node connections to the interface cell (maximum of 3).
KPLC	(3, number of child boundary cells)	Array that maps the index of the child interface cell to the corresponding parent layer number. The index is defined by looping through the child interface cells in the order of columns, rows, and layers.
IPLC	(3, number of child boundary cells)	Array that maps the index of the child interface cell to the corresponding parent row number. The index is defined by looping through the child interface cells in the order of columns, rows, and layers.
JPLC	(3, number of child boundary cells)	Array that maps the index of the child interface cell to the corresponding parent column number. The index is defined by looping through the child interface cells in the order of columns, rows, and layers.
IFACEGN	(3, number of child boundary cells)	The child cell face number (using MODPATH conventions) which is connected to a ghost node.
ICBOUND	NCOL,NROW,NLAY (child model)	A copy of the child IBOUND array.
GNHEAD	(3, number of child boundary cells)	Head at a ghost node connected to a child cell.
DHGN	(3, number of child boundary cells)	Difference between ghost-node head and parent head in the same cell.
GNFLUX	(3, number of child boundary cells)	Flux between a ghost node and a child cell.
GNFLUXR	(3, number of child boundary cells)	Relaxed flux between a ghost node and a child cell.
GNFLUXOLD	(3, number of child boundary cells)	Flux between a ghost node and a child cell from the previous LGR iteration.
HOLDC	NCOL,NROW,NLAY (child model)	Head in the child grid at the previous LGR iteration.
GNCOND	(3, number of child boundary cells)	Conductance between a ghost node and a child cell.
VCB	4	Volumetric budget values for the interface specified-head cells of the child grid: (1)—Inflow rate for current time step. (2)—Outflow rate for current time step. (3)—Cumulative volume of inflow. (4)—Cumulative volume of outflow.
HK	NCOL,NROW,NLAY (child model)	Horizontal hydraulic conductivity when using BCF.
VK	NCOL,NROW,NLAY (child model)	Estimate of the vertical hydraulic conductivity of a cell when using BCF or HUF.

Description of LGR2 Subroutines

Listed below are the subroutines that are within the gwf2lgr2 f file. Subroutines that are called from main are in bold, underlined text.

GETNAMFILLGR—Reads the Name files from the LGR2 control file.

GWF2LGR2AR—Allocates and reads data for LGR. Calls SGWF2LGR2PSV to save pointer arrays for LGR data.

GWF2LGR2DA—De-allocates LGR2 module data.

GWF2LGR2RP—Finds the mapping between the column, row, and layer of interface child cell to the corresponding location of the parent grid. Calls SGWF2LGR2PNT to change pointers for LGR2 data to the appropriate grid.

GWF2LGR2INITP—Zeros out the interior cells of the parent grid which are completely covered by the child cells. Called after the first full parent solution.

GWF2LGR2PFM—Subtract ghost-node conductance terms from the diagonal accumulation term of the matrix equations, HCOF, (if using method 2) and add ghost-node flux terms to RHS of parent grid. Calls SGWF2LGR2GNCOND to calculate ghost-node conductance.

GWF2LGR2CFM—Subtract ghost-node conductance terms from HCOF and RHS of child grid. Calls SGWF2LGR2GN-COND to calculate ghost-node conductance.

GWF2LGR2DARCY—Initialize the child heads using parent heads on the first iteration if ISHFLG=1. Calculates ghost-node heads using Darcy-planar interpolation. Relax the head change at the ghost nodes and find the location and value of the maximum head change.

GWF2LGR2FMBF—Calculates child interface ghost-node fluxes. Relax the flux change and find the location and value of the maximum flux change.

GWF2LGR2CNVG—Checks for convergence of the LGR2 iterations. Calls SGWF2LGR2PNT to change pointers for LGR data to the appropriate grid.

GWF2LGR2PBD—Calculates the volumetric budget of the parent interface ghost-node flux and writes values to a compact budget file. Calls SGWF2LGR2BFHPOT to output parent interface boundary heads and fluxes for use with the BFH2 Package.

GWF2LGR2CBD—Calculates the volumetric budget of the child interface ghost-node heads and writes associated flux values to a compact budget file. Calls SGWF2LGR2BFHCOT to output child interface boundary heads and fluxes for use with the BFH2 Package

GWF2LGR2COT—Print the parent and child ghost-node fluxes and percent differences to the screen, if requested, and the output file.

Table 5-2.   Variable in Fortran module GWFBFHMODULE.

Variable name	Size	Description
ISCHILD	Scalar	Flag: -1=parent grid, 1=child grid
IBOTFLG	Scalar	Flag: 0=bottom layer is not refined, 1=bottom layer is refined
IBFLG	Scalar	Unit number used in child IBOUND array that denotes the interface boundary
NCPP	Scalar	Refinement ratio along rows and columns
NPBNODES	Scalar	Number of interface boundary nodes in the parent grid
NCBNODES	Scalar	Number of interface boundary nodes in the child grid
NTIMES	Scalar	Number of times the interface boundary conditions are saved
NPLBEG	Scalar	Layer in the parent grid where refinement begins
NPRBEG	Scalar	Row in the parent grid where refinement begins
NPCBEG	Scalar	Column in the parent grid where refinement begins
NPLEND	Scalar	Layer in the parent grid where refinement ends
NPREND	Scalar	Row in the parent grid where refinement ends
NPCEND	Scalar	Column in the parent grid where refinement ends
IUBC	Scalar	Flag and unit number If IUBC $\neq$ 0, it is the unit number of the file where the complementary boundary conditions are saved
NGRIDS	Scalar	Number of grids simulated
BTEXT	C*17	Name of the boundary condition
IBPFLG	Number of grids -1	A list of IBOUND flags (IBFLG) used within this grid
IBB	NCBNODES	Value of IBFLG for the current grid
NPINDX	NCBNODES	Array that maps an index of child interface cells to an index of the parent interface cells based on looping through the interface cells in the order of columns, rows, and layers
KLAY	NCBNODES	Array that maps an index of child interface boundary cell to the corresponding layer number of the child The index is defined by looping over the interface cells in the order of columns, rows, and layers
IROW	NCBNODES	Array that maps an index of child interface boundary cell to the corresponding row number of the child The index is defined by looping over the interface cells in the order of columns, rows, and layers
JCOL	NCBNODES	Array that maps an index of child interface boundary cell to the corresponding column number of the child The index is defined by looping over the interface cells in the order of columns, rows, and layers
IFACEGN	(3, number of child boundary cells)	The child cell face number (using MODPATH conventions) which is connected to a ghost node
KPLAY	NCBNODES	Array that maps the index of the child interface cell to the corresponding parent layer number The index is defined by looping through the child interface cells in the order of columns, rows, and layers
IPROW	NCBNODES	Array that maps the index of the child interface cell to the corresponding parent row number The index is defined by looping through the child interface cells in the order of columns, rows, and layers
JPCOL	NCBNODES	Array that maps the index of the child interface cell to the corresponding parent column number The index is defined by looping through the child interface cells in the order of columns, rows, and layers
BFLUX	NCBNODES	Ghost-node flux across the parent-child interface Used as the coupling boundary condition for the parent grid
BFLUXCHK	NCBNODES	Ghost-node flux across the parent-child interface Used as the complementary boundary condition for the child grid
BHEAD	NCBNODES	Ghost-node head at the child interface boundary Used as the coupling boundary condition for the child grid
BCOND	NCBNODES	Ghost-node conductance at the child interface boundary Used to connect ghost-node head to the child boundary
BHEADCHK	NPBNODES	Head at the parent interface boundary Used as the complementary boundary condition for the parent grid
VCB	4	Volumetric budget values for the interface specified heads of the child grid: (1)—Inflow rate for current time step (2)—Outflow rate for current time step (3)—Cumulative volume of inflow (4)—Cumulative volume of outflow

Description of BFH2 Subroutines

Listed below are the subroutines that are within the gwf2bfh2.f file. Subroutines that are called from main are in bold, underlined text.

GWF2BFH2AR—Allocates and reads data for the BFH2 Package. Calls SGWF2BFH2PSV to save pointer arrays for BFH2 data.

GWF2BFH2DA—De-allocates BFH2 module data.

GWF2BFH2RP—Reads the column, row, and layer indices for cells along the interface where the boundary conditions are applied. Calls SGWF2LGR2PNT to change pointers for BFH2 data to the appropriate grid. Zeros out the interior cells of the parent grid that will be completely covered by the child grid.

GWF2BFH2AD—Reads the coupling boundary condition data for the current time step and also the complementary boundary condition data, if used.

GWF2BFH2FM—For child grids, calls SGWF2BFH2FMCBF to adjust HCOF and RHS with ghost-node heads and conductances along the parent-child boundary interface. For the parent grid, calls SGWF2BFH2FMPBF to apply the ghost-node flux to RHS of the parent interface boundary.

GWF2BFH2BD—Calculates the volumetric budget for the parent interface boundary specified fluxes and the child interface boundary specified heads. Calls SGWF2BFH2PNT to change pointers for BFH2 data to the appropriate grid. For the child grid, calls SGWF2BFH2CBF to find budgets for ghost-node heads at the child interface boundary. Writes interface fluxes to a compact budget file, if requested.

GWF2BFH1OT—This subroutine calls SGWF2BFH2CBD to output to the listing file a separate budget for the child interface boundary specified heads when global budgets are printed. If complementary boundary conditions are used for either the parent or child grid, then report the location and value of the maximum difference.

Appendix 6.   Relative Advantages of Ghost-Node versus Shared-Node Coupling

Advantages of the Ghost-Node Method

1.   The ghost-node method shares cell interfaces rather than nodes. In terms of aligning the grids, this is more conventional and perhaps easier for users to conceptualize.

2.   Does not require halving conductances or storages, which makes the code simpler.

3.   Does not require adjustment of stresses at the grid interface (the shared node coupling method loses the representation of stresses at the child specified-head cells, so these stresses need adjustment by the user as discussed by Mehl and Hill, 2005, p. 5–7).

4.   More flexibility in refining. The ghost-node method allows odd or even refinement ratios horizontally. Odd and even refinement ratios are allowed vertically—that is, odd and even refinement ratios can be mixed when refining vertically (the shared node method allows odd refinement ratios only).

5.   More flexible in terms of formulating the coupling equations. Two options are investigated in this report. Solving the entire system simultaneously is also possible, but is not considered in this report.

6.   Allows drying/re-wetting of interface cells (when shared nodes go dry, which are simulated as constant-head cell goes dry, the simulation stops).

7.   Allows grid refinement that can extend laterally to the edge of the parent grid (shared-node grids cannot start or end in the first or last row or column of the parent model). For example, the entire top layer of the parent model can be refined horizontally.

Advantages of the Shared-Node Method

1.   More accurate interpolation scheme (cage-shell interpolation) for three-dimensional heterogeneous systems and all unconfined and transient simulations. Steady-state, confined simulations of three-dimensional homogeneous or two-dimensional heterogeneous systems produce interpolated heads of equivalent accuracy as the interpolation scheme used in the ghost-node method.

2.   Efficient use of solvers and existing allocated memory for cage-shell interpolation with no difficulties supporting BCF, LPF, and HUF.

3.   Particle tracking using MODPATH-LGR is compatible with shared nodes and takes advantage of half cells at the interface when moving particles across grids.